高校专门用途英语（ESP）系列教材

ENGLISH *for* ENERGY HUMANITIES
Viewing and Listening

能源人文英语视听教程

总主编　赵秀凤

主　编　蔡　坤

副主编　李　音　杜芳芳
　　　　李子男

U0187556

清华大学出版社
北京

内 容 简 介

    本教材包括八个单元，内容涉及能源人文英语的八个话题，分别为能源与生活、能源与科技、能源与道德、能源与可持续发展、能源与文明、能源与人文、能源与环境、能源与未来。每个单元设有 Listening 和 Viewing 两个版块，视听练习内容丰富、形式多样，音视频资源原汁原味、语音纯正，着重培养学生的学术英语听力综合理解能力，提升学生的思辨能力和自主学习能力。

    本教材适合高等院校能源类专业本科生和研究生使用，也可供广大学术英语爱好者及相关领域从业人员阅读参考。

图书在版编目（CIP）数据

能源人文英语视听教程 / 赵秀凤主编. —北京：清华大学出版社，2024.4
高校专门用途英语（ESP）系列教材
ISBN 978-7-302-58489-6

Ⅰ. ①能… Ⅱ. ①赵… Ⅲ. ①能源人文英语—英语—听说教学—高等学校—教材
Ⅳ. ① TK01

中国版本图书馆 CIP 数据核字（2021）第 121372 号

责任编辑：周　航
封面设计：子　一
责任校对：王凤芝
责任印制：丛怀宇

出版发行：清华大学出版社
        网　　址：https://www.tup.com.cn, https://www.wqxuetang.com
        地　　址：北京清华大学学研大厦 A 座　　邮　编：100084
        社 总 机：010-83470000　　　　　　　　邮　购：010-62786544
        投稿与读者服务：010-62776969, c-service@tup.tsinghua.edu.cn
        质量反馈：010-62772015, zhiliang@tup.tsinghua.edu.cn
印 装 者：三河市春园印刷有限公司
经　　销：全国新华书店
开　　本：170mm×230mm　　印　张：12　　字　数：176 千字
版　　次：2024 年 5 月第 1 版　　　　印　次：2024 年 5 月第 1 次印刷
定　　价：59.00 元

产品编号：090351-01

前　言

　　能源是工业文明的基础，是人类文明进步的动力。能源塑造着人类的历史、现状和未来。正如莱斯利·怀特（Leslie White）所说，人类的历史是不断寻求利用新能源资源的历史。现代社会在政治、生态、文化、信仰等方面面临着诸多挑战，追根溯源，答案可能都藏在"能源"之中。因此，能源问题已经成为国际政治、经济、环境保护等诸多领域的中心议题，如气候变化、低碳经济、环境保护、可持续发展等无不与能源相关——这些话题不仅涉及国际格局和国家发展，也与民众的生活息息相关。在"百年未有之大变局"的时代背景之下，我们更加需要以一种全新的眼光审视能源议题。

　　能源人文学就是这样一种适用的理论视角。能源人文学通过批评和分析能源在历史、哲学、文化、文学、艺术等领域的书写再现，敦促人们重新审视化石能源对现代文明形态的塑造、对气候变化等"人类世"现象的直接影响，呼吁人们摆脱能源现实主义、能源技术主义及新自由主义迷思，在认识论和本体论意义上重构人与能源资源之间的关系，构想可持续发展的人类未来。能源人文从新的维度审视能源之于人的主体性、之于人类社会和人类未来的作用和功能，既弥补了人文学中能源缺席的遗憾，又填补了能源科学研究中人文性缺失的空白，具有极其重要的理论价值和实践指导意义。能源人文契合人文主义的价值和范式，体现了人文研究者的使命和担当，有助于全面认识能源的本质，把人类对能源的认识从技术主义、现实主义和物质主义的迷思中匡救回来，构建更美好的人类社会。

　　中国石油大学（北京）外国语学院自 2012 年起在全国率先成立了中国国际能源舆情研究中心，并创建了能源人文研究团队，致力于在

能源科学技术和人文社会科学之间架起一道桥梁，充分利用中外语言优势，关注、构建和传播能源人文思想；致力于引领和推动社会各界尤其是新时代大学生从人文社会科学的角度审视能源问题，特别是化石能源的相关问题，为应对生态灾难，重新思考人类发展与能源之间的关系，重塑社会文化制度、政治格局、价值理念，推动人类社会的可持续发展提供一个崭新视角。从能源人文视角出发，学习、了解和思考能源相关的热点话题，审视人类文明和国际国内社会生活，培养大学生的国际视野、人文情怀和可持续发展理念，是能源人文英语系列教材的立意所在。

《能源人文英语视听教程》内容主要涉及能源与日常生活、能源与技术、能源与伦理、能源与可持续发展、能源与文明、能源与人文、能源与环境、能源与未来，共八个单元。每个单元设有 Listening 和 Viewing 两个版块，Listening 版块包括两段音频输入材料，Viewing 版块包括两段较长的视频输入材料。视听素材选自有影响力的专业英文网站，语言真实地道。在类型上，视听素材以新闻、访谈、专题报道为主。

本教材以能源与人类福祉相关内容为依托，精心设计视听活动并配以讲解，着重培养学生的学术英语听力综合理解能力，提升学生的思辨能力和自主学习能力，帮助学生顺利使用英语学习相关领域的专业课程。

本教材具有以下特点：

1. 选材兼具思政性和真实性

本教材体现了"立德树人"的教育理念，以视听材料为载体，展现世界各国在能源领域的前沿技术，特别是中国在节能减排领域的正确立场和研究成果，引导学生更好地践行社会主义核心价值观、学习中国学者求实务新的科研精神。所有视听材料均选自国内外权威平台或知名网站，素材真实、语言地道，具有较强的实用性，能够帮助学生拓宽视野，提升综合能力。

2. 目标兼具人文性和工具性

本教材旨在培养学生的能源国际视野、学术思辨能力和英语听力能力。选材聚焦当前能源热点话题，在保证知识性和思辨性的基础上，不涉及过于专业的能源科技知识，引导学生以开阔的国际视野和人文情怀理解和思考相关能源议题。就听力技能而言，本教材从选材到练习，侧重培养学生学术英语的听力综合理解能力，包括听懂能源领域的短篇学术讲座，将大意或重点记录下来并写出小结，就讲座中没有听清楚的内容进行提问。概言之，内容上的人文性和技能上的工具性融于一体是本教材的突出特点之一。

3. 活动兼具阶梯性和多样性

本教材听力活动由简到难，通过听前预测、听时记录和听后总结，帮助学生逐渐理解大意及重点细节。听力活动具体包括单项选择、多项选择、判断正误、信息配对、排列顺序、短文填空、简短回答等多种形式。按照循序渐进的阶梯性学习规律，采用多样化的方式提高学生英语交际能力是本教材遵循的基本原则。

本教材可以作为高校本科石油、煤炭、电力、矿业、水利等能源相关专业的大学英语教材，也可以作为全国能源类相关企事业单位的专业培训教材，还可以作为从事能源政治、能源经济、能源国际关系、能源治理等领域相关工作的企业管理者或研究人员的学习材料。

本教材编写团队来自中国石油大学（北京）外国语学院。赵秀凤担任总主编，蔡坤担任主编，李音、杜芳芳、李子男担任副主编，采用交叉审校程序校对每个单元。赵秀凤负责本书的整体策划和内容审定，蔡坤编写第 1—8 单元 Viewing 部分，杜芳芳编写第 1—3 单元 Listening 部分，李音编写第 4—6 单元 Listening 部分，李子男编写第 7—8 单元 Listening 部分。由于编者水平有限，书中难免存在疏漏之处，请广大读者不吝指正。

编者

2023 年 12 月

目　录

Unit 5　Energy and Civilization............. 49

Unit 6　Energy and Humanity 61

Unit 7　Energy and Environment 71

UNIT 1
Energy and Daily Life

Lead-in

The world depends on a great deal of its energy in the form of fossil fuels. Examples of fuels include gasoline, coal and natural gas. Each day, people bathe, cook, clean, do laundry and drive using various types of fuels. However, fossil fuels are a double-edged sword. On the one hand, energy consumption has improved our lifestyles substantially. On the other hand, carbon footprint has increased so sharply that air pollution is threatening the planet. How can we create a cleaner and more efficient world? Have you heard about any advanced technology or invented products in the energy field?

Energy Talk

We will thoroughly advance the energy revolution. Coal will be used in a cleaner and more efficient way, and greater efforts will be made to explore and develop petroleum and natural gas, discover more untapped reserves, and increase production. We will speed up the planning and development of a system for new energy sources, properly balance hydropower development and ecological conservation, and develop nuclear power in an active, safe, and orderly manner. We will strengthen our systems for energy production, supply, storage, and marketing to ensure energy security.

—Xi Jinping

The 20th National Congress of the Communist Party of China

Oct. 16th, 2022

深入推进能源革命，加强煤炭清洁高效利用，加大油气资源勘探开发和增储上产力度，加快规划建设新型能源体系，统筹水电开发和生态保护，积极安全有序发展核电，加强能源产供储销体系建设，确保能源安全。

——习近平

中国共产党第二十次全国代表大会

2022 年 10 月 16 日

Listening

 ## Fuels Used in Our Daily Life

Words and Expressions			
backhoe	/ˈbækhəʊ/	n.	反铲挖土机
calcium	/ˈkælsɪəm/	n.	钙
convert	/kənˈvɜːt/	v.	（使）转变，（使）转换
crane	/kreɪn/	n.	起重机，吊车
crude oil deposits			原油矿床
crumble	/ˈkrʌmbl/	v.	（使）破碎
extract	/ɪkˈstrækt/	v.	提取，提炼
for good			永远，永久
furnace	/ˈfɜːnɪs/	n.	熔炉
radioactive decay			放射性衰变
sluice	/sluːs/	n.	水闸，闸门
transmit	/trænzˈmɪt/	v.	传送，输送
uranium	/juˈreɪnɪəm/	n.	铀

1. Listen to the recording and then match the fuels in the left column with the roles they play in daily life in the right column.

_____ 1) gasoline	A. It is the primary fossil fuel for power plants to generate electricity.
_____ 2) natural gas	B. It is consumed by nuclear power plants to generate energy.

_____ 3) coal	C. It is essential for transportation. Cars, buses and trucks run on it.
_____ 4) alcohol	D. It is called the fuel of life. It can also generate power for homes in areas near running streams and rivers.
_____ 5) uranium	E. It heats the Earth, fuels the water cycle and helps plants grow.
_____ 6) water	F. It powers the heating and cooking systems, such as stove tops, water heaters and dryers in your home.
_____ 7) solar energy	G. It is mixed with gasoline for much of the U.S. needs of liquid fuel.

2. **Listen to the recording again and fill in the blanks with what you hear.**

Gasoline

The most 1) _____ used in daily life runs cars, school buses and trucks. Gasoline and diesel are non-renewable fuels created from 2) _____ in the ground or beneath the oceans. Lawnmowers and other 3) _____ also run on gasoline. Construction sites power backhoes, dump trucks, cranes and other equipment with diesel.

Natural gas

Natural gas can power the heating systems, stove tops, water heaters and dryers in your home. Natural gas burns very cleanly and 4) _____ when burning, according to the

Natural Gas.org. This type of fuel 5) _____ _____ methane but can contain other gases as well. Natural gas often occurs as underground pockets near oil deposits. Oil 6) _____ that rise to the higher levels of underground pockets of oil trapped within rock layers. Wells tap into these pockets to remove the natural gas for use in your home.

Coal

Many electrical plants burn coal as the primary fossil fuel for powering the electrical supply for homes across the country. According to the American Coal Foundation, 7) _____ _____ fuels the electrical needs for more than half of all U.S. homes. Machines crumble the coal into small particles that get placed inside a furnace. The coal gets burned to heat water and 8) _____ that fuels a turbine to create mechanical energy. This mechanical energy converts to electrical energy in a generator then gets transmitted through substations that deliver electricity to customers.

Uranium

Although uranium isn't "burned" to make heat like coal or natural gas is, it still counts as fuel as nuclear power plants consume it and 9) _____ from it. It is also like coal or other fuels in that it is non-renewable: When the supply is used up, it is gone for good. Unlike fossil fuels, uranium creates heat through radioactive decay, a process that, weight for weight, can 10) _____. The downsides of uranium include dangerous radioactivity and waste that remains radioactive for thousands of years.

How to Reduce Our Carbon Footprint

Words and Expressions

compulsion	/kəm'pʌlʃn/	n.	强迫，强制
engulf	/ɪn'gʌlf/	v.	吞没，淹没
justifiably	/ˌdʒʌstɪ'faɪəbli/	adv.	言之有理地，无可非议地
lumbering	/'lʌmbərɪŋ/	adj.	步态笨拙的
mitigate	/'mɪtɪgeɪt/	v.	减轻，缓和
myriad	/'mɪriəd/	n.	无数，大量
relegate	/'relɪgeɪt/	v.	使贬职；使降级
spate	/speɪt/	n.	一连串，接二连三
the stick and the carrot			大棒加胡萝卜，软硬兼施
urbanite	/'ɜːbənaɪt/	n.	城市居民
viable	/'vaɪəbl/	adj.	可实施的，切实可行的

1. **Listen to the recording and choose the best answer to each of the following questions.**

1) According to the article, which of the following statements about the air pollution in China is NOT true?

   A. It occurs in many cities in China.

   B. It occurs more frequently in northern China than in southern China.

   C. It is caused by the consumption of large amounts of energy in our daily life.

2) Why is it hard for individuals to take the required measures to curb air pollution according to the author?

A. People usually put their personal concerns in the first place.

B. There is no element of compulsion or positive incentives in the implementation of these measures.

C. Both A and B.

3) According to the author, a greener lifestyle includes the following behaviors EXCEPT _____.

A. people switch on electrical appliances in their home when necessary

B. people make a car journey wherever they go

C. people use sharing bikes

4) Which of the following statements about the way the Personal Carbon Trading system works is NOT true?

A. Those whose carbon footprint remains below the average level would be rewarded in some way.

B. Those with excessive use of energy would not pay for it.

C. Each citizen is issued with a certain number of points every year.

5) The best title for this article may be "_____".

A. Measures Taken in the United Kingdom to Curb Air Pollution

B. How to Reduce Carbon Footprint in China

C. Air Pollution in China

2. Listen to the recording again and fill in the blanks with what you hear.

Air Pollution

The recent spate of pollution to 1) _____ Beijing—and many other cities in China—should lead us to the conclusion as to what measures we can take as 2) _____ _____ who are faced with this growing problem on a daily basis to help mitigate it.

Economic Growth

China's 3) _____ of economic growth since the turn of the century has improved the lifestyles of many of its population in ways which would have been 4) _____ _____ just a matter of a few decades previously.

Increase of Carbon Emission

But this has also led to the development of lifestyles which consume amounts of energy that would have been 5) _____ _____, with the carbon footprint of the average Chinese urbanite increasing from the mere imprint of a baby's footstep to one which would 6) _____ crashing about before us.

So the next time we complain or 7) _____ feel concerned about the dangerous levels of air pollution, we should also look ourselves squarely in the face and consider what we should do to help bring our own carbon footprint down by a few sizes.

Before you make your next car journey, you ought to consider whether there is any other viable 8) _____, or the next time you reach to switch on one of the myriad of electrical appliances in your home, you ought to consider whether it is absolutely necessary for you to do so.

Viewing

 ## Solar Energy Products You Should Buy

Video One		

Words and Expressions		
Bluetooth	/'blu:tu:θ/	n. 蓝牙（一种无线传输技术）
hybrid	/'haɪbrɪd/	n. 混合物
infrastructure	/'ɪnfrəstrʌktʃə(r)/	n. 基础设施
patented	/'pætntɪd/	adj. 得到……专利权的
photovoltaic	/ˌfəʊtəʊvɒl'teɪɪk/	adj. 光电池的
plug	/plʌg/	v. 填塞
scenario	/sə'nɑːrɪəʊ/	n. 可能发生的情况

1. Watch a video clip about solar energy products and decide whether the statements are true or false. Write "T" for true and "F" for false.

_____ 1) HELIOS is a hybrid Bluetooth headphone that captures solar power.

_____ 2) With a full battery, HELIOS can deliver half an hour of music.

_____ 3) If you're out of HELIOS battery, you will run out of music.

_____ 4) The Kali PAK, a green energy generator, fits dozens of off-the-grid energy needs.

_____ 5) The 39-amp battery is powerful enough to charge your iPhone up to 40 times, an iPad up to 120 times.

2. **Watch the video clip again and answer the following questions.**

1) What is the function of the mini USB port equipped in HELIOS?

_____

2) How can HELIOS allow you to answer calls hands-free?

_____

3) When do we always run out of power according to the speaker?

_____

4) Why is Kali PAK completely sustainable and autonomous?

_____

5) Who will Kali PAK be donated to? Why?

_____

Video Two

Words and Expressions

compatible	/kəmˈpætəbl/	adj.	兼容的；可共存的
deplete	/dɪˈpliːt/	v.	消耗，耗尽
faucet	/ˈfɔːsɪt/	n.	水龙头

irrigation	/ˌɪrɪˈɡeɪʃən/	n.	灌溉
profile	/ˈprəʊfaɪl/	n.	简介；外形
reserve	/rɪˈzɜːv/	n./v.	储备

1. Watch a video clip on ODO and put a tick before the features of the domestic irrigation solution system.

_____ 1) advanced

_____ 2) efficient

_____ 3) sustainable

_____ 4) smart

_____ 5) expensive

2. Watch a video clip again and fill in the blanks with what you hear.

ODO knows the water needs of your plants and automatically adjusts the irrigation 1) _____ to weather forecasts and to soil conditions. This flexibility improves efficiency and water saving. It allows you to save more than 2) _____ liters of potable water each year. It is completely self-sufficient and sustainable. Our technology for data 3) _____ is the core of the product. It works at a low frequency that allows a wider range key and low battery consumption. ODO is connected to the 4) _____ and constantly monitors the weather forecasts from the web. It also compares these data for a double-check with its embedded weather station. It's plug-and-play and 5) _____ with every existing irrigation system. ODO takes care of your plants for you and you can easily control the entire process. It has an extensive plants 6) _____ to control your cultivation through its sensors. You can always know the health status of your plants. The ODO 7) _____ is the perfect tool to monitor and manage your system wherever you are. You just need an Internet connection. ODO is a social movement for water sustainability too: You can share your experiences and your successes on popular social 8) _____ or on its dedicated community. Now you can open your faucets, being sure not to waste water. We are offering an affordable price to make this 9) _____ possible. We want to spread ODO all over the world, but we need your support. The more ODO is spread 10) _____, the more affordable will be our product and we'll make a larger revolutionary impact on our planet.

UNIT 2
Energy and Technology

Lead-in

American researchers have experimented with a new energy storage technology—looking to sodium to make better batteries in the future. Compared to most commonly used lithium-ion batteries, sodium batteries are less costly but more powerful. Another good example of energy technology is Songdo in South Korea, a city which is becoming smarter and more energy efficient with the help of smart technology. In the video clip, ten energy efficiency tips for households are discussed in detail. Can you come up with more ideas about energy efficient households?

Energy Talk

We must revolutionize energy technology, and upgrade the related industrial structure. We should encourage innovation in technology, in industry, and in business models, and pursue green and low-carbon energy development suited to our national conditions and adapted to positive international trends in the energy technology revolution. We will combine such innovation with new and high technology in other fields, and transform our energy technology and related industries into a new powerhouse to drive the overall industrial upgrading of our country.

—Xi Jinping

The Sixth Meeting of the Central Leading Group
on Financial and Economic Affairs

June 13th, 2014

推动能源技术革命，带动产业升级。立足我国国情，紧跟国际能源技术革命新趋势，以绿色低碳为方向，分类推动技术创新、产业创新、商业模式创新，并同其他领域高新技术紧密结合，把能源技术及其关联产业培育成带动我国产业升级的新增长点。

——习近平

中央财经领导小组第六次会议

2014 年 6 月 13 日

Listening

Researchers Look to Sodium to Make Better Batteries

Words and Expressions

cost-effective		*adj.*	划算的；成本效益好的
electron	/ɪˈlektrɒn/	*n.*	电子
energy density			能量密度
ion	/ˈaɪən/	*n.*	离子
lithium	/ˈlɪθɪəm/	*n.*	锂
property	/ˈprɒpəti/	*n.*	（物质、物体的）特性，属性
sodium	/ˈsəʊdɪəm/	*n.*	钠
storage container			储存容器

1. Listen to the recording and choose the best answer to each of the following questions.

1) The best title for this article may be "_____".

   A. Experiments on Better Batteries for Electric Vehicles

   B. Advertisement of Batteries That Are More Powerful

   C. Researches on a New Generation of Batteries

2) Which of the following statements about sodium battery is NOT true?

   A. It has already been launched in the market.

   B. It has different properties from that of lithium battery.

    C. The experiments of sodium batteries get financial support
from the government.

3) The advantages of sodium battery include the following
EXCEPT _____.

    A. it is much more abundant in supply

    B. it would cost less to collect

    C. it is easily found in the water

4) Which of the following statements about lithium battery is
NOT true?

    A. It powers everything from smartphones to computers to
electric vehicles.

    B. It is costly and not easy to collect.

    C. The energy density of lithium battery has no room for
improvement.

5) The experiments of the sodium batteries by the University of
California are financially supported by _____.

    A. Environmental Protection Agency

    B. the U.S. National Science Foundation

    C. United States Department of Energy

2. **Listen to the recording again and fill in the blanks with what
you hear.**

Sodium Battery

Scientists have long 1) _____ to make
batteries that are more powerful, but cost less to build. In the

United States, researchers are experimenting with sodium to see whether it can 2) _____ batteries in the future. Sodium is a soft, 3) _____. It is plentiful and found in seawater. The most common battery used today is made of lithium ion. These batteries power everything from smartphones to computers to electric vehicles. Researchers from the University of California, San Diego, are attempting to 4) _____ of batteries powered by sodium instead of lithium. The U.S. National Science Foundation is providing financial support for the experiments.

Lithium Battery vs. Sodium Battery

Shirley Meng is a member of the research team. "At a society level, I think people really think that a battery is 5) _____ _____—like it's an old subject." But Meng says the process of developing better batteries is still 6) _____ _____. In fact, she says the energy density of batteries in use today "can still be doubled or tripled". The California researchers are studying lithium ion batteries, but in the next few years plan to begin 7) _____. Team member Hayley Hirsh says she looks forward to working more with sodium development in the future. "We want to use sodium instead of lithium because it has different properties. And also, sodium is much more abundant." Lithium is costly and not easy to collect because it is widely spread across many parts of the world. Large amounts of water and energy are also 8) _____.

Smart Cities

1. Listen to the recording and decide whether the following statements are true or false. Write "T" for true and "F" for false.

_____ 1) Smart cities are just unrealistic science-fiction.

_____ 2) IBM has developed the software to predict traffic congestion.

_____ 3) 65% of the world's population is expected to live in cities by 2050.

_____ 4) The designer of the city Songdo is the company IBM.

_____ 5) Smart technology is expected to make cities greener, more sustainable, and more efficient.

2. **Listen to the recording again and answer the following questions.**

1)   What are the facilities like in a smart city?

_____

2)   What will residents of Songdo be able to use their "telepresence" systems for?

_____

3)   What are the technology firms mentioned in the article in the development of software to solve city problems?

_____

4)   How does the technology firm IBM develop software to predict traffic jams?

_____

5)   What does Dan Hill think the most important thing in smart cities will be?

_____

Viewing

Top 10 Energy Efficiency Tips for Your Home

Words and Expressions

attic	/ˈætɪk/	n.	阁楼，顶楼
audit	/ˈɔːdɪt/	n.	审计
basement	/ˈbeɪsmənt/	n.	地下室
foundation	/faʊnˈdeɪʃn/	n.	地基
gigajoules	/ˌdʒɪɡeɪˈʒuːlz/	n.	十亿焦耳
real estate			不动产
renovation	/ˌrenəˈveɪʃn/	n.	翻新，修复

1. Watch a video clip about energy efficiency and decide whether the statements are true or false. Write "T" for true and "F" for false.

_____ 1) The EnerGuide Homeowner Information Sheet shows the energy rating of Brian's 1956 home.

_____ 2) The home constructed in the 1950s is using 80% of its energy on space heating.

_____ 3) The home constructed in the 1950s is using 20% of its energy on lights appliances.

_____ 4) Brian's 1956 home uses 236 gigajoules of natural gas and electricity per year.

_____ 5) If Brian's home were built to today's standards, it would consume less than a quarter of 236 gigajoules of natural gas and electricity per year.

2. **Watch the video clip again and fill in the blanks with what you hear.**

Brian's 1956 home uses most of its energy to 1) _____ _____ in the home. Not surprisingly, the 2) _____ _____ focused on things that will save energy used to heat the home. This is another useful tool that we can look at where the home is 3) _____. In this instance, this home is losing 29% of its heat through 4) _____, another 25% of it through 5) _____ and the rest is made up to attic, main walls, 6) _____.

Video Two

Words and Expressions

aspirate	/ˈæspəˌreɪt/	v.	吸气
furnace	/ˈfɜːnɪs/	n.	熔炉，火炉
incandescent	/ˌɪnkænˈdesnt/	adj.	白热的；白炽的
induction	/ɪnˈdʌkʃn/	n.	（电或磁的）感应
insulation	/ˌɪnsjuˈleɪʃn/	n.	隔热材料；绝缘
leakage	/ˈliːkɪdʒ/	n.	泄漏；渗漏物
phantom	/ˈfæntəm/	n.	幻影
thermostat	/ˈθɜːməstæt/	n.	恒温器

1. Watch a video clip and rank the energy efficiency measures in the order of importance.

_____ use the mighty furnace

_____ do a home energy assessment

_____ add the smart thermostat

_____ seal up the building envelope

_____ redo the windows

_____ add insulation

_____ change the water heater

_____ unplug devices or use a smart power bar

_____ improve appliances

_____ change lights to LEDs

2. Watch the video clip again and fill in the blanks with what you hear.

So there you have it, our top 10 list of energy efficiency tips for your home. Remember, every home is different. But by checking these things in your home you will find 1) _____ that can make a big difference in your home. Brian Finley now knows where his home can be improved. 2) _____ can make a big difference. But when it comes to energy efficiency, you just might want to 3) _____ as well. We've seen homes where owners have reduced their energy consumption by 4) _____ by simple measures, such as changing lights to LEDs, adding smart thermostats, unplugging the monster beer fridge, and taming their energy phantoms. As we've learned on 5) _____, it's always a good idea to make your home energy-efficient first. But having done that, you just might want to add 6) _____ to your home and produce your own energy. You'll save money on electricity over the life of the system and you'll be producing your own clean energy 7) _____ for the next 25 years. As a bonus, you won't 8) _____ .

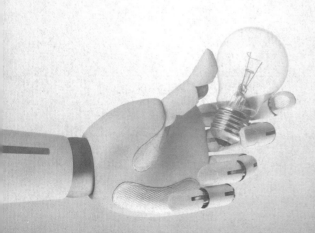

UNIT 3

Energy and Ethics

Lead-in

Energy ethics raise our concerns about energy-related issues including energy availability, climate change and energy security. The Power African Initiative is an energy program, aiming to improve energy access in African countries south of the Sahara Desert. The 2015 Paris Accord calls for countries to cut the production of carbon dioxide and other gases linked to rising temperatures. The ethics of nuclear energy are also worthy of our attention, because nuclear energy has proven to be both beneficial and disastrous. At a time of growing energy demand and rapid climate change, what should we do to balance our energy demands with our concerns for anthropogenic climate change? And what should we do to create a better energy future for us all?

Energy Talk

First, we must revolutionize energy consumption, and rein in irrational energy use. We need to impose strict controls on overall energy use, effectively implement a policy in which energy conservation is the top priority, and save energy across the board in all spheres of economic and social activity. We should also adjust the structure of the energy industry, make energy conservation a priority in urbanization, and encourage an attitude to consumption characterized by diligence and thrift. We must build an energy-conserving society.

—Xi Jinping
The Sixth Meeting of the Central Leading Group
on Financial and Economic Affairs
June 13th, 2014

推动能源消费革命，抑制不合理能源消费。坚决控制能源消费总量，有效落实节能优先方针，把节能贯穿于经济社会发展全过程和各领域，坚定调整产业结构，高度重视城镇化节能，树立勤俭节约的消费观，加快形成能源节约型社会。

——习近平
中央财经领导小组第六次会议
2014 年 6 月 13 日

Listening

Providing Electricity to Poor Communities in Kenya

Words and Expressions

availability	/əˌveɪləˈbɪləti/	n.	可得性，可用性
foundation	/faʊnˈdeɪʃn/	n.	基金会
General Electric			美国通用电气公司
hack	/hæk/	v.	非法侵入
human waste			人体排泄物
illegally	/ɪˈliːɡəli/	adv.	非法地
initiative	/ɪˈnɪʃətɪv/	n.	倡议，措施
launch a campaign			发起一项运动
slum	/slʌm/	n.	贫民窟

1. Listen to the recording and choose the best answer to each of the following questions.

1) The goal of the campaign called Power Africa Initiative is to
_____.

A. increase power access in all African countries

B. increase power access in African countries south of the Sahara Desert

C. increase power access in African countries north of the Sahara Desert

2) Which of the following statements about the project called Afrisol Energy is NOT true?

   A. It generates electricity by burning coal.

   B. It can generate electricity enough to power the local school and neighborhood.

   C. It is financed by the Power Africa Initiative.

3) The following statements about the use of electricity in the local school in the past are true EXCEPT _____.

   A. they decided to live without electricity because of poor access

   B. illegal connections to the school's power supply are very serious

   C. hacking of their power system leads to higher cost for the school

4) From what the mother Beatrice Onchan'ga says, we can conclude that _____.

   A. the school is now still lack of reliable power supplies

   B. children in the school are now able to depend on reliable power supplies

   C. children cannot come to school that early and leave that late because the school is very dark

5) What is the percentage of Kenyans who have access to electricity according to the World Bank?

   A. 32%.

   B. 26%.

   C. 23%.

2. **Listening to recording again and fill in the blanks with what you hear.**

Power Africa Initiative

In 2013, President Barack Obama 1) _____ called the Power Africa Initiative. Its goal is to increase the 2) _____ in African countries south of the Sahara Desert. Millions of people there are unable to depend on reliable power supplies. The American-supported program is providing money for a number of projects, including one that creates electricity from 3) _____.

Afrisol Energy

"Mukuru Kwa Njenga" is the name of a community close to Nairobi where about 100,000 people live. Many of them are poor. Until recently, most did not have electricity. Those who did had 4) _____ to power lines. Amos Nguru had the idea for Afrisol Energy, a project that produces electricity from human waste. Two years ago, Mr. Nguru received an 5) _____ from the Power Africa Initiative. The money came from General Electric and the United States African Development Foundation. The project now produces 15 kilowatts of electricity. That is enough to power a nearby school and serve the local neighborhood. Mr. Nguru says his project meets the needs of the community.

Illegal Connection in the Local School

Deborah Mwandagina is deputy head teacher of the local primary school. She says in the past that there were too many illegal connections to her school's power supply. She says this resulted in higher costs for the school, so it decided to stop using electricity. She says, "We have been having 6) _____ and time and again they 7) _____ and they connect illegally to their homes. We have been having such kind of challenges so we just decided we cannot live with electricity. It can't do us any good, so we 8) _____."

U.S. Withdraws from the Paris Agreement

Words and Expressions			
administration	/ədˌmɪnɪˈstreɪʃn/	n.	管理部门，行政部门
carbon dioxide			二氧化碳
Celsius	/ˈselsɪəs/	n.	摄氏
drought	/draʊt/	n.	久旱，旱灾
opponent	/əˈpəʊnənt/	n.	对手，竞争者
Paris Accord			巴黎协议
Paris Agreement			巴黎协定
treaty	/ˈtriːti/	n.	条约，协定
withdraw	/wɪðˈdrɔː/	v.	撤回；撤离

1. **Listen to the recording and decide whether the following statements are true or false. Write "T" for true and "F" for false.**

_____ 1) The Paris Accord is an agreement approved by leaders from the European Union to limit climate change.

_____ 2) The United States has made official announcement to withdraw from the treaty.

_____ 3) The United States' withdrawal from the treaty may not last long in view of the attitude of Joe Biden, Trump's main opponent in the presidential election, on this issue.

_____ 4) The agreement aims to limit the increase in average temperatures worldwide to "well below" 1.5 degrees Celsius.

_____ 5) The Paris Accord imposes certain targets on countries for cutting down emission of greenhouse gases.

_____ 6) According to the report, the United States is the world's biggest producer of heat-trapping gases.

_____ 7) The Trump administration refuses to carry out measures made by the federal government to cut greenhouse gases.

_____ 8) Despite the United States' withdrawal from the Paris Agreement, U.S. states, cities and businesses are still committed to the cutting down on emission of greenhouse gases.

2. **Listen to the recording again and answer the following questions.**

1) What did Joe Biden, the Democratic Party's candidate for president, promise to do if he is elected?

_____

2) What could be the disastrous effect caused by any temperature increase greater than 2 degrees Celsius?

_____

3) In recent weeks, which countries have joined together in setting national targets to stop emitting more greenhouse gases into the atmosphere?

_____

4) What is Biden's reaction to calls for the United States to return to the Paris Accord?

_____

5) What is Germany's government's comment on the United States' withdrawal from the Accord?

_____

Viewing

The Ethics of Nuclear Energy

Video One

Words and Expressions

dissipate	/ˈdɪsɪpeɪt/	v.	驱散，使消散
humanity	/hjuːˈmænəti/	n.	人性；人道
justify	/ˈdʒʌstɪfaɪ/	v.	证明……有理，为……辩护
lethal	/ˈliːθl/	adj.	致命的；破坏性的
meddle	/ˈmedl/	v.	干涉，插手
milligram	/ˈmɪlɪɡræm/	n.	毫克（千分之一克）
plutonium	/pluːˈtəʊnɪəm/	n.	［化］钚

pro and con			争论；利与弊
radioactive	/ˌreɪdɪəʊˈæktɪv/	*adj.*	放射性的
radioactivity	/ˌreɪdɪəʊækˈtɪvəti/	*n.*	放射性；辐射能
reactor	/rɪˈæktə(r)/	*n.*	反应器；［核］反应堆
untapped	/ʌnˈtæpt/	*adj.*	未开发的，未利用的

1. **Watch a video clip and tick the advantages of nuclear energy mentioned.**

_____ **1)** It is a new energy.

_____ **2)** It is good for the environment compared to fossil fuels.

_____ **3)** It has saved many lives.

_____ **4)** Most of the technology is up to date.

2. **Watch the video clip again and answer the following questions.**

1) Why is nuclear energy really good for the environment?

_____

2) How will a ton of radioactive nuclear waste be usually dealt with?

_____

3) According to a 2013 NASA study, what has happened since 1976 due to the use of nuclear energy?

_____

4) How have people been affected by the reduction to the amount of fossil fuels in the air?

_____

Video Two

Words and Expressions

thyroid	/ˈθaɪrɔɪd/	n.	甲状腺
redress	/rɪˈdres/	n.	赔偿；矫正
property	/ˈprɒpəti/	n.	财产；特性
atomic	/əˈtɒmɪk/	adj.	原子的；原子能的
proliferation	/prəˌlɪfəˈreɪʃn/	n.	激增
controversial	/ˌkɒntrəˈvɜːʃl/	adj.	有争议的
byproduct	/ˈbaɪˌprɒdʌkt/	n.	副产品

1. **Watch a video clip and decide whether the statements are true or false. Write "T" for true and "F" for false.**

_____ 1) The most recent nuclear emergency was Fukushima, more formally known as the Fukushima Daiichi disaster.

_____ 2) An energy accident happened at the Fukushima nuclear power plant on March 16th, 2017.

_____ 3) No one died from the major radiation leak, though 37 people were injured.

_____ 4) Two workers were hospitalized due to radiation burns.

_____ 5) The outcome of this accident was a rise in lung cancer among the people living near the power plant.

2. **Watch the video clip again and fill in the blanks with what you hear.**

First, let me say the seven major nuclear disasters that have happened around the world in the past 1) _____ years. Though three of these at ordeals were fairly self-contained, the other four actually render entire parts of countries unfit for human 2) _____. Another issue with nuclear energy is nuclear power plant waste and pollution. Like I mentioned earlier, it is healthier for the environment but only in 3) _____ to fossil fuels. Last, plutonium is a large threat to us as humans. For one, just a milligram of plutonium could kill a human being and a kilogram is all you need to make an 4) _____ bomb. America is one of 190 parties to have the Nuclear Non-Proliferation Act, which is part of the 5) _____ on the Non-Proliferation of Nuclear Weapons that was created in 1968. The Nuclear Non-Proliferation Act aims for, and I quote, "more effective international control over the transfer and use of nuclear materials equipment and nuclear technology for 6) _____ purposes to prevent proliferation".

In the context of this, nuclear energy can be an extremely 7) _____ talking point. If you were to look at it morally and ethically, it doesn't seem to be hurting many people yet and could provide better electrical opportunities for many places in the world. But one of the 8) _____ of nuclear energy is the creation of very powerful nuclear weapons. And for this reason and this alone, it is hard to agree with nuclear energy from a moral 9) _____. This is also the stance that I take as an individual concerning nuclear energy. I think that until we know for certain that we can control the effects of a nuclear 10) _____ and what people use them for, it is not a good idea to encourage the use of nuclear energy or nuclear power plants.

UNIT 4

Energy and Sustainability

Lead-in

Society desires to achieve sustainable development. Energy consumption must be coordinated with population, economy, and environment to achieve sustainable social development. Traditional energy sources such as coal, oil and gas have been shown to emit harmful greenhouse gases that contribute to global warming. These fossil fuels are also limited in quantity and are not infinite. Sustainable energy is renewable energy that does not emit carbon dioxide and other greenhouse gases. It is renewable as it harnesses the power of the planet, meaning we will not run out of these natural sources of energy in the future. Do you know the various sources of sustainable energy? How do they contribute to promoting the transition to green and low-carbon development and building a clean and beautiful world?

Energy Talk

We need to pursue inclusive and sustainable development. Earth is the only home for humanity. We must follow a people-centered approach, foster a sound environment to buttress sustainable economic and social development worldwide, and achieve green growth. China attaches great importance to addressing climate change. We will strive to peak carbon dioxide emissions before 2030 and achieve carbon neutrality before 2060.

China supports APEC in advancing cooperation on sustainable development, improving the List of Environmental Goods, and making energy more efficient, clean and diverse. We need to enhance economic and technological cooperation, promote inclusive trade and investment, support the development of small- and medium-sized enterprises, scale up support for women and other vulnerable groups, share experience on eliminating absolute poverty and strive to deliver the 2030 Agenda for Sustainable Development.

—Xi Jinping

The APEC Informal Economic Leaders' Retreat

July 16th, 2021

　　坚持包容可持续发展。地球是人类赖以生存的唯一家园。我们要坚持以人为本，让良好生态环境成为全球经济社会可持续发展的重要支撑，实现绿色增长。中方高度重视应对气候变化，将力争 2030 年前实现碳达峰、2060 年前实现碳中和。

　　中方支持亚太经合组织开展可持续发展合作，完善环境产品降税清单，推动能源向高效、清洁、多元化发展。我们要加强经济技术合作，促进包容性贸易投资，支持中小企业发展，加大对妇女等弱势群体的扶持力度，分享消除绝对贫困的经验，努力落实 2030 年可持续发展议程。

——习近平

亚太经合组织领导人非正式会议

2021 年 7 月 16 日

Listening

How to Increase Your Renewable Energy Use?

Words and Expressions

discerning	/dɪˈsɜːnɪŋ/	adj.	有辨识能力的
ethanol fuel			乙醇燃料
harness	/ˈhɑːnɪs/	v.	治理；利用
implementation	/ˌɪmplɪmenˈteɪʃn/	n.	实施，执行
legislator	/ˈledʒɪsleɪtə(r)/	n.	立法者
receptor	/rɪˈseptə(r)/	n.	感受器
replenish	/rɪˈplenɪʃ/	v.	补充，装满
skyrocket	/ˈskaɪrɒkɪt/	v.	猛涨
solar panel			太阳能板
supplemental	/ˌsʌpləˈmentl/	adj.	补充的
wind turbine			风力涡轮机

1. **Listen to the recording and decide whether the statements are true or false. Write "T" for true and "F" for false.**

_____ 1) Solar panels are very expensive at present and as demand continues to skyrocket, the prices are soaring dramatically.

_____ 2) Ethanol fuel source is comprised of corn alcohol and gasoline and helps lower the carbon emissions that are produced when you drive.

_____ 3) Wind energy is fully renewable, sustainable and stable.

_____ 4) Renewable energy is not only unlimited in supply, but also environmentally friendly.

_____ 5) Companies should be encouraged to use sustainable energy in the making of their products.

2. **Listen to the recording again and fill in the blanks with what you hear.**

1) Renewable energy involves the _____, _____ and _____ of sustainable resources. This means that the source from which the energy is cultivated can be replenished and reused over and over again.

2) One way you can increase your renewable energy use is by having _____ installed on your home or office building.

3) Putting _____ in your gas tank instead of _____ is another way to increase your renewable energy use.

4) Wind energy is another growing renewable energy source. You can take advantage of wind power by purchasing a small _____ and installing it at your home.

5) Renewable energy sources will continue to grow in _____ across the globe, although it may take some time before options are _____ in all areas.

Making Flying Green and Sustainable

Words and Expressions

aerodynamic	/ˌeərəʊdaɪˈnæmɪk/	adj.	空气动力学的
aviation	/ˌeɪvɪˈeɪʃn/	n.	航空
biofuel	/ˈbaɪəʊfjuːəl/	n.	生物燃料
carbon neutral			碳中和
grounded	/ˈɡraʊndɪd/	adj.	（人）不出门的；（飞机）停飞的
hybrid	/ˈhaɪbrɪd/	n.	混合动力
minimize	/ˈmɪnɪmaɪz/	v.	最小化
offset	/ˈɒfset/	n.	抵消；补偿

1. **Listen to the recording and answer the following questions.**

1) Why is the number of people taking to the skies increasing?

_____

2) What was the hot topic at the recent Dubai Air Show?

_____

3) What are the benefits of using biofuel to power aircraft?

_____

4) What ways are aircraft manufacturers constantly looking at to make their planes more fuel-efficient?

_____

2. Listen to the recording again and fill in the blanks with what you hear.

1)   Jumping on a plane and jetting off on holiday or a business trip is the _____ for many of us. But while air travel is costing us less, the cost to the _____ is going up.

2)   _____ is something we're all aware of now, and aviation companies know that some of the blame for this is being pointed at them. Recent developments have focused on reducing _____ airlines burn.

3)   While we could think twice about taking a flight in the first place, _____ might be the answer to reducing emissions and minimizing the environmental damage.

4)   Being _____ is the ultimate goal for the aviation industry, and one British airline, EasyJet, has recently said it would become the world's first major net zero carbon airline by offsetting carbon emissions.

Viewing

Innovative Companies: Sustainable Energy

Words and Expressions

circular loop			圆形环
electrolyzer	/ɪˈlektrəʊlaɪzə/	n.	电解剂
pivotal	/ˈpɪvətl/	adj.	关键的
refine	/rɪˈfaɪn/	v.	提炼；改善
mindful	/ˈmaɪndfəl/	adj.	留心的，注意的
sequestration	/ˌsiːkwəˈstreɪʃn/	n.	吸收

1. Watch a video clip and match the figures in the left column with the facts in the right column.

1) 21%	A. the estimated percentage of electricity production in the EU from variable renewables by 2014
2) 7%	B. the actual percentage of global electricity production from variable renewables in 2018
3) 39%	C. the predicted percentage of global electricity production from variable renewables by 2040

| 4) $6.1 billion | D. the market for battery energy storage systems by 2023 |
| 5) $13.1 billion | E. the market for battery energy storage systems in 2018 |

2. **Watch the video clip again and fill in the blanks with what you hear.**

Mindful of the high levels of CO_2 in indoor air, the company plans to use its technology to clean what we breathe in our offices by 1) _____ its technology in buildings. It says that this will improve people's well-being and make them more 2) _____ all while producing valuable clean resources with the CO_2 pollution.

First, we have here city air. You always have a ventilation in the building and you push it through the ventilation unit through our carbon 3) _____ unit, and then you are able to get lower CO_2 air indoors, which will be the 4) _____ of people's well-being. Then we have the electrolyzer. We break down water into hydrogen and oxygen and then we have a 5) _____ unit which can do the fuel part or hydrocarbons. And if this building would be connected to a gas grid, you could provide synthetic 6) _____ which you can pump into the gas grid. So the gas grid could be added as an energy storage, or we could have a car 7) _____ station that you can fill up the car tank.

The main aim is to use the hydrocarbons to fuel vehicles where they can be mixed with fossil fuels. And any CO_2 emissions 8) _____ from the engine can then be pumped back into the solar tier process, closing the 9) _____ loop. The product can also be refined to produce a cleaner gas to heat and power homes. Scientists often wish they could pluck solutions for a sustainable future out of 10) _____ air. The solar tier team might just come close.

Video Two

Words and Expressions

preserve	/prɪˈzɜːv/	v.	保护；保持
equivalent	/ɪˈkwɪvələnt/	adj.	相等的
utilization	/ˌjuːtəlaɪˈzeɪʃn/	n.	使用，利用
influx	/ˈɪnflʌks/	n.	流入，注入
curtailment	/kɜːˈteɪlmənt/	n.	缩减；缩短
hydrogen	/ˈhaɪdrədʒən/	n.	氢
kilowatt-hour			[电]千瓦时
one-size-fits-all			通用的；一刀切

1. **Watch a video clip and decide whether the statements are true or false. Write "T" for true and "F" for false.**

_____ 1) In Wales there's a company taking the concept of zero emissions up.

_____ 2) The hydrogen-powered vehicles can travel up to 250 miles on 1.5 kilograms of hydrogen—that's the equivalent of nearly 300 miles per gallon.

_____ 3) According to the research of China's five-year projects, by 2030 more than 35% of the energy demand will be supplied by the renewables.

_____ **4)** By 2040 60% to 70% of the energy demand will be supplied by the renewables.

2. **Watch the video clip again and answer the following questions.**

1) What is the multi-energy system?

It's a site that coordinates the _____ of energy, especially renewable energy across different energy _____.

2) How do multiple energy systems work?

They increase the efficiency and flexibility of both energy _____, so they can be used when and where they're needed, and _____ as opposed to getting tailed and wasted.

3) What is the strong seasonal pattern of the hydropower?

In a flooded season, there is too much water to produce electricity that the load is not enough. Part of the gas system can use the energy surplus to produce a lot of gas using the hydropower. The gas can be used in _____ or even to produce electricity in the dry season so that it can provide _____.

UNIT 5

Energy and Civilization

Lead-in

All human advances have been connected to advances in producing and using energy from earth's natural resources. Energy is a historically significant element in civilization. The evolution of human civilization can be seen as the history of energy development, from the primitive period to the industrial period and finally to this highly technological era today. Do you know how the exploitation of clean energy has opened a new era of energy utilization and led us to the prosperity of the world economy and an unprecedented civilization in human history?

Energy Talk

The Chinese civilization has always valued harmony between man and Nature as well as observance of the laws of Nature. It has been our constant pursuit that man and Nature could live in harmony with each other. Ecological advancement and conservation have been written into China's Constitution and incorporated into China's overall plan for building socialism with Chinese characteristics. China will follow the Thought on Ecological Civilization and implement the new development philosophy. We will aim to achieve greener economic and social development in all aspects, with a special focus on developing green and low-carbon energy. We will continue to prioritize ecological conservation and pursue a green and low-carbon path to development.

—Xi Jinping

The Leaders Summit on Climate

April 22nd, 2021

中华文明历来崇尚天人合一、道法自然，追求人与自然和谐共生。中国将生态文明理念和生态文明建设写入《中华人民共和国宪法》，纳入中国特色社会主义总体布局。中国以生态文明思想为指导，贯彻新发展理念，以经济社会发展全面绿色转型为引领，以能源绿色低碳发展为关键，坚持走生态优先、绿色低碳的发展道路。

——习近平

领导人气候峰会

2021 年 4 月 22 日

Listening

India Plugs into Low-Cost Solar Technology

<table>
<tr><td colspan="4">Words and Expressions</td></tr>
<tr><td>demystify</td><td>/diːˈmɪstɪfaɪ/</td><td>v.</td><td>揭秘；阐明</td></tr>
<tr><td>excel</td><td>/ɪkˈsel/</td><td>v.</td><td>超过</td></tr>
<tr><td>extol</td><td>/ɪkˈstəʊl/</td><td>v.</td><td>赞美</td></tr>
<tr><td>grassroots</td><td>/ˈɡraːsˌruːts/</td><td>adj.</td><td>基层的；乡村的</td></tr>
<tr><td>parabolic mirror</td><td></td><td></td><td>抛物柱面镜</td></tr>
<tr><td>solder</td><td>/ˈsəʊldə(r)/</td><td>v.</td><td>焊接</td></tr>
<tr><td>sprocket</td><td>/ˈsprɒkɪt/</td><td>n.</td><td>链轮齿</td></tr>
</table>

1. Listen to the recording and decide whether the statements are true or false. Write "T" for true and "F" for false.

_____ 1) India seems to be good at making things smaller and cheaper.

_____ 2) The $250 computer and the $35 TV are just two of India's latest innovations.

_____ 3) Nearly one third of India's rural population—more than 300 million people has either no electricity or just a few hours of it a day.

_____ 4) Chonzom was chosen by her community to come to the workshop to learn about solar technology.

_____ 5) Barefoot College aims at helping India's rural poor by teaching them to make and install low-cost solar panels.

_____ 6) Sanjit Bunker Roy who started Barefoot College 25 years ago is among *Time* magazine's top 10 most influential people for 2010.

_____ 7) Some of the women at Barefoot College helped design and build solar cooker.

_____ 8) Roy says his program merely provides the space for women in rural areas to develop creativity and cooperation.

2. **Listen to the recording again and fill in the blanks with what you hear.**

1) In this sunlit workshop, Tenzing Chonzom solders parts onto a device that regulates electrical currents. It will eventually be connected to a _____, allowing it to power everything from _____ to _____.

2) Chonzom is 50 years old, and one of two dozen people being trained here as _____. Most have had no _____.

3) "You'll find it everywhere in India, this _____ to be able to improvise and fix things without having gone through any formal education," Roy added. "They have this _____ that we haven't been able to define, appreciate or respect yet."

4) Sita Devi says she wanted to make a _____ with materials that are easily available, even in _____. She says the cooker saves time—and _____—by reducing the need for women to wander outside the village in search of _____ for cooking.

5) Roy says his program can provide the space for women to develop _____.

Shenzhen's Silent Revolution

Words and Expressions

behemoth	/bɪ'hiːmɒθ/	n.	巨兽
charging pile			充电桩
confine	/kən'faɪn/	v.	限制
depot	/'depəʊ/	n.	车站；停车场
diesel	/'diːzl/	n.	柴油
flume	/fluːm/	n.	车队
hiss	/hɪs/	n.	嘶嘶声
megacity	/'megə‚sɪti/	n.	大都市
nitrogen oxides			氧化氮；氮氧化合物
non-methane hydrocarbon			非甲烷碳氢化合物
piercing	/'pɪəsɪŋ/	adj.	刺耳的

1. Listen to the recording and choose the best answer to each of the following questions.

1) Which of the following is not the benefit from the switch from diesel buses to electric?

A. Less noise pollution.

B. Less traffic jams.

C. Reduction in CO_2 emissions.

D. Cuts in pollutants.

2) To keep Shenzhen's electric vehicle fleet running, the city has built around _____ charging piles.

A. 60,000

B. 440,000

C. 160,000

D. 40,000

3) Why was Shenzhen able to go all-electric?

A. Donation from large companies.

B. Subsidies from central and local government.

C. Subsidies from the UN.

D. Donation from celebrities.

4) This fast-growing megacity of 12 million is also expected to achieve cuts in pollutants such as _____.

A. nitrogen oxides

B. non-methane hydrocarbons

C. particulate matter

D. all of the above

2. **Listen to the recording again and fill in the blanks with what you hear.**

1) Shenzhen now has _____ electric buses in total and is noticeably quieter for it. Shenzhen Bus Group estimates it has been able to conserve _____ tonnes of coal per year and reduce annual CO_2 emissions by _____ tonnes. Its fuel bill has _____.

2) To keep Shenzhen's electric vehicle fleet running, the city has built around 40,000 _____.

3) Shenzhen Bus Company has 180 depots with their own _____ installed.

4) China's drive to reduce the choking smog that envelops many of its major cities has propelled _____.

5) Cities such as London and New York are _____ their drive towards electric buses. London plans to make all single-decker buses _____ by 2020, and all double-deckers _____ by 2019. New York plans to make its bus fleet _____ by 2040.

Viewing

What Happens When You Change Energy Culture?

Video One

Words and Expressions

aspiration	/ˌæspəˈreɪʃn/	n.	抱负
captain	/ˈkæptɪn/	n.	船长；队长
circumstance	/ˈsɜːkəmstəns/	n.	环境，情况，情形
destiny	/ˈdestəni/	n.	命运
habitual	/həˈbɪtʃuəl/	adj.	习惯的，惯常的

intervention	/ˌɪntəˈvenʃn/	n.	介入，干涉
mobility	/məʊˈbɪləti/	n.	流动性，移动性
norm	/nɔːm/	n.	社会准则，行为规范
vacuum	/ˈvækjuːm/	n.	真空；空白
victim	/ˈvɪktɪm/	n.	受害者，遇难者

1. **Watch a video clip and decide whether the statements are true or false. Write "T" for true and "F" for false.**

_____ 1) Up till around 2004 we saw a rapid increase in the total vehicle kilometers traveled.

_____ 2) There's also a marked increase in the amount of people who drive when they reach 65 or when they retire.

_____ 3) A lot of New Zealanders actually aspire for active transport.

_____ 4) A lot of New Zealanders do always walk, cycle or use public transport.

_____ 5) We are not the victim to circumstances but rather the captain of our own destiny.

2. **Watch the video clip again and answer the following questions.**

1) What are transport agencies going to do traditionally if vehicle kilometers traveled is going to continue up?

_____

2) What are the two possible reasons for driving less and less?

_____

3) What are the three things that all interact with each other to build up patterns of behavior?

_____

4) What is the mismatch between aspirations and actions in New Zealand?

_____

5) How can New Zealanders deal with the mismatch?

_____

Video Two

Words and Expressions

combustion engine			内燃机
complicated	/ˈkɒmplɪkeɪtɪd/	adj.	复杂的；难处理的
mechanism	/ˈmekənɪzəm/	n.	机制；机能
minister	/ˈmɪnɪstə(r)/	n.	部长
passionate	/ˈpæʃənət/	adj.	热烈的；激昂的
regulate	/ˈregjuleɪt/	v.	控制，管理
sophisticated	/səˈfɪstɪkeɪtɪd/	adj.	复杂的，精密的
stability	/stəˈbɪləti/	n.	稳定（性）；稳固
trade-off			交易；权衡

 1. Watch a video clip and list the four different visions of the future of the transport system.

1) _____

2) _____

3) _____

4) _____

2. **Watch the video clip again and fill in the blanks with what you hear.**

We valued the research community immensely. We are here to provide advice to ministers on issues. There are often views which were expressed, where people are very 1) _____ about an issue and they will want change, but there are always 2) _____. So, when you're providing advice, one of the key things to have is 3) _____ to allow you to provide advice which will deliver the best 4) _____ for New Zealand. So, the energy culture's framework helps us think about a really 5) _____ situation in a relatively simple way. If people do start changing the energy culture, whether it is because of outside influences or because they are actually changing the 6) _____, wanting to do things quite differently, like "Generation Y" around mobility, then, when people start changing those energy cultures, their "starts" create a 7) _____ change. Those collective changes can actually start to influence 8) _____ and can start to change the way companies run their businesses; they can start to change the sorts of infrastructure that governments need to provide; they start to change, maybe, some of the policy 9) _____ that their governments have. So, that framework, although it's very simple, invites us to think about all of the different parts of this very complex system, and think about both what locks us into 10) _____ or habit, and also what it is that drives change.

UNIT 6

Energy and Humanity

Lead-in

Energy has a great impact on humanity issues, such as poverty, health, gender equality, employment, education, etc. Have you ever thought about the harmful effects the lack of modern energy could have on human beings? How could modern energy make a difference on people's life especially in developing countries?

Energy Talk

China will firmly uphold the common interests of the world. China will take an active part in cooperation within the United Nations, the WTO, the G20, APEC, Shanghai Cooperation Organization and other institutions, and promote more discussions on such issues as trade and investment, the digital economy, and green and low-carbon development. China will support the fair distribution of and unimpeded trade in vaccines and other key medical supplies across the world. China will promote high-quality Belt and Road cooperation so that more countries and peoples will benefit from its development opportunities and real outcomes. China will actively join in global efforts to tackle climate change and safeguard food and energy security, and provide more assistance to fellow developing countries within the framework of South-South cooperation.

—Xi Jinping

The Opening Ceremony of the Fourth China International Import Expo

Nov. 4th, 2021

中国将坚定不移维护世界共同利益。中国将积极参与联合国、世界贸易组织、二十国集团、亚太经合组织、上海合作组织等机制合作，推动加强贸易和投资、数字经济、绿色低碳等领域议题探讨。中国将支持疫苗等关键医疗物资在全球范围内公平分配和贸易畅通。中国将推动高质量共建"一带一路"，使更多国家和人民获得发展机遇和实惠。中国将积极参与应对气候变化、维护全球粮食安全和能源安全，在南南合作框架内继续向其他发展中国家提供更多援助。

——习近平

第四届中国国际进口博览会开幕式

2021 年 11 月 4 日

Listening

Solar Energy Makes the Difference in Africa

Words and Expressions

affordable	/əˈfɔːdəbl/	adj.	负担得起的
coordinate	/kəʊˈɔːdɪneɪt/	v.	协调
generator	/ˈdʒenəreɪtə(r)/	n.	发电机
grant	/grɑːnt/	n.	资助
installment	/ɪnˈstɔːlmənt/	n.	安装
kerosene	/ˈkerəsiːn/	n.	煤油
sub-Saharan		n.	撒哈拉以南地区

1. Listen to the recording and decide whether the statements are true or false. Write "T" for true and "F" for false.

_____ 1) The majority of people living in sub-Saharan Africa lack access to electricity.

_____ 2) Only a few people living in sub-Saharan Africa spend significant amounts of their income on costly and unhealthy forms of energy.

_____ 3) The Scaling Off-Grid Energy Grand Challenge for Development focuses on pay-as-you-go solar home systems.

_____ 4) The goal of this Challenge is to provide 20 million off-grid households in sub-Saharan Africa with clean, affordable electricity by 2050.

_____ 5) The cost of solar technology is rising fast therefore most people in sub-Saharan Africa cannot afford it.

_____ 6) People can purchase the solar power-generating equipment they need and pay for it in daily installments from their telephones.

2. **Listen to the recording again and fill in the blanks with what you hear.**

1) Lack of _____ is one of the chief obstacles to growth and development in sub-Saharan Africa.

2) But if there is one thing Africa is not lacking, it is _____. And that means a lot; indeed, for some of the world's poorest people, it _____ in the world.

3) So far, the Challenge has made 40-plus _____ in early-stage off-grid energy companies, which are expected to result in some 4.8 million new _____.

4) In these newly-electrified communities, businesses can flourish, clinics can safely store vaccines, and students may study long after dark. Indeed, _____ can enable entire communities to escape the cycle of extreme poverty.

5) Thanks to _____, many people are getting electricity for the first time, every day.

Lack of Access to Energy

Words and Expressions			
in a nutshell			简而言之
pandemics	/pænˈdemɪks/	*n.*	流行病
prioritize	/praɪˈprətaɪz/	*v.*	优先处理，优先考虑
transition	/trænˈzɪʃn/	*n.*	转换；转型
vaccine	/ˈvæksiːn/	*n.*	疫苗

1. Listen to the recording and answer the following questions.

1) What problems are caused by burning carbon fuels?

_____

2) What are the consequences to lack of access to energy?

_____

3) What should governments do in order to accelerate the transition to an affordable, reliable, and sustainable energy system?

_____

4) What suggestion is given to reduce carbon emissions?

_____

2. Listen to the talk again and fill in the blanks with what you hear.

1) In a nutshell, without _____, countries will not be able to power their economies.

2) Nearly _____ now have access to electricity, but reaching the unserved 789 million around the world—548 million people in sub-Saharan Africa alone—that lack access will require increased efforts.

3) Without electricity, women and girls have to _____, clinics cannot _____ for children, many schoolchildren cannot _____ at night, and people cannot _____.

4) Lack of access to energy may hamper efforts to contain COVID-19 across many parts of the world. _____ are key to preventing disease and fighting pandemics—from _____ and _____ for essential hygiene, to enabling communications and IT services that connect people while maintaining social distancing.

5) Businesses can _____ and commit to sourcing 100% of operational electricity needs from _____.

Viewing

Humans and Energy

Video One

Words and Expressions

appetite	/ˈæpɪtaɪt/	n.	胃口，食欲
converter	/kənˈvɜːtə(r)/	n.	变换器；变频器

domestication	/dəˌmestɪˈkeɪʃn/	n.	驯服，驯养
dominate	/ˈdɒmɪneɪt/	v.	在……中占首要地位；控制
nutritious	/njuˈtrɪʃəs/	adj.	有营养的
penetrate	/ˈpenətreɪt/	v.	穿透；进入
permanent	/ˈpɜːmənənt/	adj.	永久的；不断出现的
pyramid	/ˈpɪrəmɪd/	n.	金字塔
steam engine			蒸汽机
unappeasable	/ˈʌnəˈpiːzəbl/	adj.	无法平息的；不能满足的

1. Watch a video clip and put a tick before the energy which comes from the sun.

_____ **1)** muscle

_____ **2)** water

_____ **3)** wind

_____ **4)** coal

_____ **5)** oil

_____ **6)** natural gas

2. Watch the video clip again and fill in the blanks with what you hear.

1) For more than _____ of human history, the main source of energy to do work was muscle, either human or animal. And the fuel for that muscle was food, usually plants, and plants _____ get their energy from the sun.

2) The first great energy technology was fire. It enabled us to cook which gave us a greater _____ of available

food and thus more fuel for our muscles. Fire also led to _____ and improvements in tools.

3) Another _____ advance in energy was the domestication of plants and animals. By domesticating plants, humans redirected the sun's energy to create more _____ and energy-producing food.

4) The only energy that we had that didn't _____ from the sun was water power. Since wind _____ comes from the sun's heating the air, sailing ships and windmills are kind of solar power.

5) Industrialization _____ new forms of fuel in coal, and later oil and natural gas. These fuels are just really, really old forms of fossilized plant and animal matter. So again, they're _____ from the sun, but we don't think of it this way.

6) England was where coal use really took off, thanks to the _____. According to Crosby, "It is the first machine to provide significantly large amounts of power not derived from muscle, water, or wind."

Video Two

Words and Expressions

boom	/bu:m/	v.	激增，繁荣
catastrophic	/ˌkætə'strɒfɪk/	adj.	灾难性的；糟糕的
contrivance	/kən'traɪvəns/	n.	发明
fallout	/'fɔːlaʊt/	n.	（核爆炸后的）放射性坠尘，核辐射；余波
gadget	/'gædʒɪt/	n.	小配件，小装置

| illumination | /ɪˌluːmɪˈneɪʃn/ | *n.* | 照明；阐明 |
| reconcile | /ˈrekənsaɪl/ | *v.* | 使和谐一致，调和 |

1. **Watch a video clip and decide whether the statements are true or false. Write "T" for true and "F" for false.**

_____ 1) By the end of the 20th century, there were a billion cars in the world, and humans were using 70 million barrels of oil each day.

_____ 2) According to Crosby, humanity's primary energy use has increased twenty times over since 1850 and nearly five times over since 1950.

_____ 3) Oil and natural gas are the most important fuels for this electricity boom, although as of 2006, 40%–50% of humans, most of them living in the tropics, still rely on wood for fuel.

_____ 4) Nuclear accidents have happened, both Windscale in England in 1957, and San Loren in France in 1969 were catastrophic.

_____ 5) Overall, nuclear power has never accounted for more than 5% of the world's energy supply.

2. **Watch the video clip again and fill in the blanks with what you hear.**

1) After steam-powered manufacturing, it was a short chronological leap to electricity, which was used to power machines, and for _____. Electric light was a really big deal because it provided a clean and _____ way to allow people to work after dark.

2) Oil was _____ because it could power not only electricity plants, ships, and trains, but also the internal combustion engine, which makes cars and trucks possible. Crosby maintains that "the internal combustion engine powering the automobile, truck, and tractor has for a _____ been the most influential contrivance on the planet".

3) The first nuclear plant providing power for homes opened in the _____ in 1954, and some countries, notably France, still rely heavily on nuclear energy. Despite initial _____ from scientists and science fiction writers, nuclear power never caught on in the U.S. partly because it's really expensive.

4) Another problem is that no one can figure out what to do with the _____ waste that nuclear energy produces. But the biggest reason nuclear power fails to catch on is that people think nuclear power is dangerous, believing that nuclear plants can easily turn into huge _____.

5) The U.S. had a nuclear scare in 1979 with an accident at Three Mile Island in Pennsylvania. Although there were no immediate _____, thousands of people in the vicinity were forced to _____, and the cleanup took years, and cost millions.

UNIT 7
Energy and Environment

Lead-in

Gases that trap heat in the atmosphere are called greenhouse gases. In addition to carbon dioxide, methane is another source of greenhouse gases, adding to the warming of our planet. To reduce the impacts of greenhouse emissions, great efforts have been made to develop renewable energy such as wind and solar energy. However, the exploitation of renewable energy brings new challenges to our attention, ranging from the rising cost of energy bills to the possible harmful effects on the sensitive plant and animal life. Do you know other challenges of harnessing renewable energy?

Energy talk

We encourage simple, moderate, green, and low-carbon ways of life, and oppose extravagance and excessive consumption. We will launch initiatives to make Party and government offices do better when it comes to conservation, and develop eco-friendly families, schools, communities, and transport services.

—Xi Jinping

The 19th National Congress of the Communist Party of China

Oct. 18th, 2017

倡导简约适度、绿色低碳的生活方式，反对奢侈浪费和不合理消费，开展创建节约型机关、绿色家庭、绿色学校、绿色社区和绿色出行等行动。

——习近平

中国共产党第十九次代表大会

2017 年 10 月 18 日

Listening

Tons of Methane Gas Could Be Trapped Under Antarctica

Words and Expressions

basin	/'beɪsn/	n.	盆地；流域
deposit	/dɪ'pɒzɪt/	n.	沉积物，沉积层
hydrate	/'haɪdreɪt/	n.	水合物
ice sheet			冰层
methane	/'mi:θeɪn/	n.	甲烷，沼气
sediment	/'sedɪmənt/	n.	沉淀物；沉积物
seep	/si:p/	n.	（油，水）渗出地表的地方
the Antarctica Circle			南极圈
the Arctic Circe			北极圈
the South Pole			南极

1. **Listen to the recording and complete each of the sentences with a choice from the box.**

A. 4,000,000,000	B. 21,000,000,000	C. 150,000
D. 35,000,000	E. 25%	F. 50%

1) The scientists say _____ of the West Antarctic ice sheet and _____ percent of the East Antarctic sheet are on sedimentary basins.

2) Some _____ years ago, Antarctica was much warmer than today and contained life forms.

3) Studies have found that there are _____ seeps in Alaska and Greenland where methane had escaped.

4) According to an international team of scientists, up to _____ tons of methane gas could be trapped under ice-covered areas of Antarctica.

5) It is said that the sedimentary basins in the Antarctic area hold about _____ tons of carbon.

2. **Listen to the recording again and fill in the blanks with what you hear.**

1) It has been found that tons of methane could be trapped under _____ areas of Antarctic.

2) Methane was formed from _____, some of which became trapped in material that had fallen to the bottom. The material at the bottom was called _____.

3) The Antarctic ice sheet covers _____, but not the _____ sea. Methane _____ are also found at the bottom of oceans.

4) The study in the journal *Nature Geoscience* showed that some of the seeps are _____ ancient methane. The source may be natural gas or coal _____ beneath the lakes.

5) Some scientists note that the possibility of a fast methane _____ could _____ rising temperatures on Earth.

Solar Plant Raises Environment Concerns

Words and Expressions			
boiler	/'bɔɪlə(r)/	n.	锅炉，汽锅；蒸汽发生缸
collide	/kə'laɪd/	v.	碰撞
deploy	/dɪ'plɔɪ/	v.	调动；利用
molten salt			熔盐
sensitive	/'sensətɪv/	adj.	敏感的；脆弱的
thermal	/'θɜːml/	adj.	热的，热量的
turbine	/'tɜːbaɪn/	n.	涡轮机

1. **Listen to the audio clip and decide whether the statements are true or false. Write "T" for true and "F" for false.**

_____ 1) A huge solar plant is being built in California.

_____ 2) Specially designed mirrors are deployed to collect solar energy, which will drive turbine to make electricity.

_____ 3) Environmentalists do not support the idea of solar power because they are concerned about the effect of power plants on sensitive environment.

_____ 4) There have been reports of birds dying or suffering at the Ivanpah Plant.

_____ 5) Bright Source Energy has already spent more than $50 million to move endangered desert tortoises away from the power plant.

2. **Listen to the recording again and fill in the blanks with what you hear.**

1) A California company is building a huge solar power plant, known as the Ivanpah _____.

2) The solar plant is one of the highest _____ of sunlight in the world.

3) "We can store the sun's thermal energy in the form of _____ _____, so we can produce electricity even when the sun goes down", explained Desmond, who works for the California company.

4) Environmentalists are concerned that the solar plant can really affect how the species and the animals and the plants are able to survive _____.

5) Some birds die after _____ with solar equipment which the animals mistake for water. Other birds were killed or suffered _____ after flying through the _____ heat at the solar thermal plant.

Viewing

 Energy and Environment

Video One

Words and Expressions

accumulate	/əˈkjuːmjəleɪt/	v.	积累，积聚
compound	/ˈkɒmpaʊnd/	n.	化合物；混合物
lithium	/ˈlɪθɪəm/	n.	锂
obsolescence	/ˌɒbsəˈlesns/	n.	废弃；陈旧过时
purifier	/ˈpjʊərɪfaɪə(r)/	n.	清洁器
rechargeable	/riːˈtʃɑːdʒəbl/	adj.	可再充电的
remediation	/rɪˌmiːdɪˈeɪʃn/	n.	补救；纠正
retail	/ˈriːteɪl/	n.	零售
sludge	/slʌdʒ/	n.	泥浆；烂泥；沉淀物
smog	/smɒg/	n.	烟雾
sodium	/ˈsəʊdɪəm/	n.	钠
toxic	/ˈtɒksɪk/	adj.	有毒的；中毒的

1. Watch a video clip and choose the best answer to each of the following questions.

1) The following topics are explored in this video EXCEPT
_____.

A. why fossil fuels our history

B. why global warming gets worse

C. how we'll provide clean energy and water for the entire planet

D. how we'll clean up the toxic waste we've accumulated

2) Why didn't you and still don't see solar as a big deal?

A. Because of high price.

B. Because of Moore's Law.

C. Because of climate change.

D. Because of environmental pollution.

3) Which is NOT true about solar?

A. In 1984, solar was $30 a kilowatt-hour.

B. By 2014, solar was down to 16 cents a kilowatt-hour.

C. By 2040, we will be producing 100% of the world's needs from solar.

D. By 2042, we will be producing 800% of the world's needs from solar.

4) The future battery has the following features EXCEPT _____.

A. cheap and quiet

B. no maintenance

C. lasting decades

D. extremely expensive

5) The almost perfect battery developed by MIT and Samsung has the following features EXCEPT _____.

A. rechargeable

B. never wears out

C. overheat

D. 30% smaller

2. **Watch the video clip again and fill in the blanks with what you hear.**

Let's transform waste into value. Take a look at the 1) _____ _____ project going underway. Launched in 2015 in Rotterdam, it's the largest air 2) _____ in the world. It's a tower twenty-three feet tall. It collects the pollution from the air and it not only 3) _____ it. It compresses it into tiny cubes that they use to make 4) _____ _____ out of. How cool is that! There are more towers planned in cities around the world. How about energy 5) _____? Well, forget about fossil fuels. That's so 20th century. It's the sun that's going to be the key 6) _____ of energy for our planet. The sun produces 10,000 times of our needs on a 7) _____ basis. So in other words, it produces way more energy than we will ever be able to use. It's not enough to produce the energy. You then have to 8) _____ it for future use. That means battery technology is 9) _____. Batteries of the future will be using sodium and water instead of lithium, 10) _____ energy evenly, and safe enough to eat.

Video Two

Words and Expressions

dung	/dʌŋ/	n.	动物的粪便
gallon	/ˈɡælən/	n.	加仑；大量液体
nutrient	/ˈnjuːtriənt/	n.	营养物，养分
pave	/peɪv/	v.	铺；为……铺平道路
pesticide	/ˈpestɪsaɪd/	n.	杀虫剂；农药
prototype	/ˈprəʊtətaɪp/	n.	原型；蓝本
recycle	/riːˈsaɪkl/	v.	回收利用；重新使用
refurbish	/riːˈfɜːbɪʃ/	v.	刷新；擦亮
respiratory	/rəˈspɪrətri/	adj.	呼吸的
sewer	/ˈsuːə(r)/	n.	下水道，污水管
sterilize	/ˈsterəlaɪz/	v.	消毒，杀菌
vertical	/ˈvɜːtɪkl/	adj.	垂直的，直立的

1. **Watch a video clip and decide whether the statements are true or false. Write "T" for true and "F" for false.**

_____ 1) There's an old steel factory in Newark, New Jersey. It's now the world's largest vertical farm.

_____ 2) The farm recycles materials. It uses a lighting system and uses less energy.

_____ 3) The farm requires only 5% as much soil as traditional farms.

_____ 4) The farm use more effective pesticide.

_____ 5) The full crop cycle is just 60 days.

2. **Watch the video clip again and answer the following questions.**

1) What is the life like in African villages?

_____

2) How do people cook and heat their huts in African villages?

_____

3) Why don't the children go to school there?

_____

4) What is the big problem of burning animal dung?

_____

5) Why is the new 21st century toilet a good choice for villagers in Africa?

_____

UNIT 8
Energy and Future

Lead-in

Islay, a small island in Scotland, has frequently been a pioneer of the testing of renewable energy technology, focusing on projects that utilize the power of the wind, the sea, the sun and the earth. Besides, Islay is the home to a great number of whisky distilleries. The byproducts taken from distilling whisky are turned into a form of alcohol, which can be used as a clean fuel in transport. Islay is a small community with only 3,500 people, but it is taking big steps to a green energy future. Have you ever heard of Islay's green energy projects? Would you like to visit the small island one day?

Energy Talk

In the face of climate change, a major challenge to all of humanity, we need to advocate green and low-carbon development, actively promote solar, wind and other sources of renewable energy, work for effective implementation of the Paris Agreement on climate change, and continue to strengthen our capacity for sustainable development.

—Xi Jinping

The Opening Ceremony of the Eighth Ministerial Conference of the Forum on China-Africa Cooperation

Nov. 29th, 2021

面对气候变化这一全人类重大挑战，我们要倡导绿色低碳理念，积极发展太阳能、风能等可再生能源，推动应对气候变化《巴黎协定》有效实施，不断增强可持续发展能力。

——习近平

中非合作论坛第八届部长级会议开幕式

2021 年 11 月 29 日

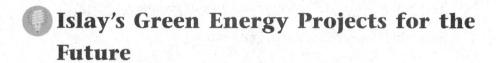

Listening

Islay's Green Energy Projects for the Future

Words and Expressions			
centerpiece	/ˈsentəpiːs/	n.	核心
distillery	/dɪˈstɪləri/	n.	酿酒厂
forge	/fɔːdʒ/	v.	稳步前进
operational	/ˌɒpəˈreɪʃənl/	adj.	操作的，运作的
respire	/rɪˈspaɪə(r)/	v.	呼吸
storm	/stɔːm/	v.	强有力地行动
the European Commission			欧盟委员会

1. Listen to the recording and choose the best answer to each of the following questions.

1) According to the passage, what is NOT true about Islay's green energy projects?

A. The island has built a wave power station.

B. The Wind Farm has been operational.

C. The island's green energy projects also include geothermal heating and solar energy.

D. The RESPIRE Project aims to show that Islay island has the potential to go totally green.

2) What does the Wind Farm aim to achieve?

   A. It will provide around half of the island's energy needs.

   B. It will achieve 100% self-sufficiency in renewable energy.

   C. It will be commercially operational.

   D. It will sponsor other energy projects.

3) What is the central part of the RESPIRE Project?

   A. Building three wind turbines.

   B. Providing electricity for wave-powered bus.

   C. Launching a community-owned company.

   D. Transforming waste hot water into heat.

4) What green energy project is NOT mentioned in the story?

   A. Wave station provides electricity for batteries of bus.

   B. A whisky distillery uses waste hot water to heat the local swimming pool.

   C. New houses are being built with geothermal heating.

   D. Nuclear energy provides half of the electricity of the island.

5) What is the population of Islay?

   A. 100 thousand.

   B. 50 thousand.

   C. 3 thousand.

   D. 1 thousand.

2. **Listen to the recording again and fill in the blanks with what you hear.**

Islay, with its green energy projects, is becoming a 1) _____ _____ to the world. 2) _____ the power of the wind, the sea, the sun and the earth is the dream of the island. It already has the world's only 3) _____ _____ operational wave power station. Its designer, Tom Heath, says the 4) _____ energy in the sea is almost 5) _____. The RESPIRE Project 6) _____ by the European Commission aims to demonstrate the 7) _____ of island communities going totally green. Islay is forging ahead with other green energy projects. One of its world-famous whiskey 8) _____ _____ uses waste hot water to heat the local swimming pool. Islay has a wave-powered bus—its batteries are 9) _____ _____ electricity produced by the wave station. New houses are being built with geothermal heating, 10) _____ warmth fifty meters below the Earth's surface. And, even in a climate where there's more rain than sun, this Gaelic Language Center is powered by solar panels.

Scottish Whisky Tested as Alternative to Fossil Fuels

Words and Expressions			
alcohol	/ˈælkəhɒl/	n.	酒精，乙醇
barley	/ˈbɑːli/	n.	大麦
bioethanol	/ˌbaɪəʊˈeθənɒl/	n.	生物乙醇；生物燃料
biofuel	/ˈbaɪəʊfjuːəl/	n.	生物燃料

distilling	/dɪsˈtɪlɪŋ/	n.	蒸馏
facility	/fəˈsɪləti/	n.	设施
grain	/greɪn/	n.	谷粒，谷物
pot ale			酒糟
rye	/raɪ/	n.	黑麦
sugar cane			甘蔗
wheat	/wiːt/	n.	小麦
whisky	/ˈwɪski/	n.	威士忌

1. **Listen to the recording and tick the statements which are true about biobutanol.**

_____ 1) Biobutanol, based on distilling whisky, can be used as fuel.

_____ 2) Biobutanol has only 70% of energy produced from gasoline.

_____ 3) In the future, biobutanol is expected to replace gasoline.

_____ 4) Automobiles have to be transformed for the biobutanol to work.

_____ 5) A center for producing biobutanol is likely to be operational in Scotland within three years.

2. **Listen to the recording again and answer the following questions.**

1) Which place is the largest producer of whisky in the world?

_____

2)   What are the raw materials used to produce whisky?

_____

3)   What is the process of making whisky?

_____

4)   What are the two byproducts of whisky that are used to create biobutanol?

_____

5)   What are the sources of bioethanol?

_____

Viewing

Hydrogen—the Fuel of the Future

Video One

Words and Expressions

fiber	/ˈfaɪbə/	n.	纤维，丝
flammable	/ˈflæməbl/	adj.	易燃的，可燃的
ignition	/ɪgˈnɪʃn/	n.	燃烧；点火，点燃
obstacle	/ˈɒbstəkl/	n.	障碍（物）
perceive	/pəˈsiːv/	v.	意识到，察觉
reliance	/rɪˈlaɪəns/	n.	依靠；依赖

1. **Watch a video clip and choose the best answer to each of the following question.**

1) How many Tesla Model 3s have been produced to date according to Bloomberg?

   A. 1,000.

   B. 2,000.

   C. 5,000.

   D. 12,000.

2) What's the current production rate of Tesla Model 3?

   A. 1,000 per week.

   B. 2,000 per week.

   C. 5,000 per week.

   D. 12,000 per week.

3) What is the relationship between supply and demand in the market of lithium-ion battery?

   A. The demand is growing faster than the supply.

   B. The demand is growing slower than the supply.

   C. The demand and supply are growing simultaneously.

   D. The demand and supply are declining simultaneously.

4) Hydrogen has three obstacles that need to overcome to become a viable energy source EXCEPT _____.

   A. safety

   B. infrastructure

   C. cost

   D. policy

2. **Watch the video clip again and fill in the blanks with what you hear.**

Obstacles	Details
Safety	Hydrogen has a relatively low ignition 1) _____ _____ and a very wide ignition range for air to fuel mixture percentages. The fact that it's pressurized makes explosions a worry, but it has one 2) _____ advantage over oil derived fuels: It's lighter than air. It can be purged using 3) _____ valves in the event of a fire and if it does ignite, it won't pull around the vehicle, engulfing it and its 4) _____ in flames.
Infrastructure	Battery-operated vehicles have had a huge head start in this space. The electric grid is a prebuilt transportation and generation 5) _____ for the fuel the battery-operated vehicles require, and installing a 6) _____ in your driveway or garage isn't a huge challenge. Hydrogen doesn't have such 7) _____ to kick-start the hydrogen economy. There are a few large scale production 8) _____ in the world, with the largest being Shell's Rhineland oil refinement facility.
Cost	Shell and ITM took the next 9) _____ step to keep cost down. They built a hydrogen production and storage facility on site. The production facility is placed just behind the main 10) _____ and is capable of producing 80 kilograms of hydrogen a day.

Video Two

Words and Expressions

conductivity	/ˌkɒndʌkˈtɪvəti/	n.	传导性；传导率；电导率
integration	/ˌɪntɪˈɡreɪʃn/	n.	结合；整合；一体化
Irish-speaking		adj.	说爱尔兰语的（人）
mainland	/ˈmeɪnlænd/	n.	大陆，本土
obscure	/əbˈskjʊə(r)/	adj.	鲜为人知的，不著名的
pipeline	/ˈpaɪplaɪn/	n.	管道，输油管道
prohibitively	/prəʊˈhɪbɪtɪvli/	adv.	过分地；过高地
well-established		adj.	已为大家接受的；根深蒂固的

1. **Watch a video clip and decide whether the statements are true or false. Write "T" for true and "F" for false.**

_____ 1) Electrolysis is a process of separating water into hydrogen and oxygen.

_____ 2) Pure water is very conductive.

_____ 3) If hydrogen has any hope of becoming a popular fuel source, we first need to get its price down to be competitive with batteries and fossil fuels.

_____ 4) The cost of hydrogen production by electrolysis is partially dependent on electricity prices.

_____ 5) This hydrogen facility at the Shell station can form an important part of the renewable grid infrastructure going forward.

2. Watch the video clip again and fill in the blanks with what
you hear.

One tiny group of 1) _____ islands
in the Bay of my home county of Galway is attempting to do
just this. The Aran Islands are 2) _____
Irish-speaking islands—popular resorts for their unique
3) _____—that would have historically depended
completely on the 4) _____ for fuel. There
are no trees here, no coal, no turf, no oil, but what they do
have in plentiful supply is 5) _____ and
wind energy. They are the perfect 6) _____ to
develop a mini hydrogen economy—an economy where they
7) _____ their own renewable energy and
8) _____ their own field to heat their homes
and power their vehicles. Who knows, these tiny obscure Irish
islands could be the 9) _____ of the world's first
self-sustained, renewable, 10) _____ hydrogen
economy.

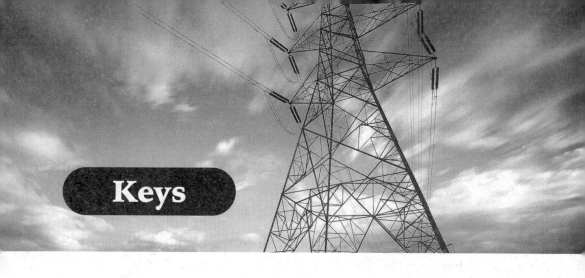

Keys

Unit 1　Energy and Daily Life

Listening

Fuels Used in Our Daily Life

1. Listen to the recording and then match the fuels in the left column with the roles they play in daily life in the right column.

   1) C　　2) F　　3) A　　4) G　　5) B　　6) D　　7) E

2. Listen to the recording and fill in the blanks with the what you hear.

   1) obvious fuel

   2) crude oil deposits

   3) maintenance equipment

   4) creates abundant energy

   5) is mostly comprised of

   6) emits gases

   7) coal-powered electricity

   8) create steam

   9) extract energy

   10) yield as much as 1 million times the energy

How to Reduce Our Carbon Footprint

1. Listen to the recording and choose the best answer to each of the following questions.

   1) B   2) C   3) B   4) B   5) B

2. Listen to the recording and fill in the blanks with what you hear.

   1) engulf

   2) individuals

   3) impressive record

   4) beyond recognition

   5) unimaginable several decades ago

   6) resemble a lumbering giant

   7) justifiably

   8) alternative

Viewing

Solar Energy Products You Should Buy

Video One

1. Watch a video clip on solar energy products and decide whether the statements are true or false. Write "T" for true and "F" for false.

   1) T   2) F   3) F   4) T   5) F

2. Watch the video clip again and answer the following questions.

   1) People can top off its battery by connecting it to a power source.

   2) Connect HELIOS to your smartphone: Turn on Bluetooth; connect to HELIOS; and listen freely to your music.

3) We always seem to run out of power, either because we choose to get out to nature and away from the grid, or in the case of a natural disaster the grid chooses to get away from us.

4) It can rely solely on its patented solar panel with a unique angle dial and comes with a rain cover, so people can keep the power going in case of rain.

5) They will be donated to African communities who lack energy infrastructure and make their life a little easier.

Video Two

1. **Watch a video clip on ODO and put a tick before the features of the domestic irrigation solution system.**

   1) √   2) √   3) √   4) √   5) \

2. **Watch a video clip again and fill in the blanks with what you hear.**

   1) profile

   2) 7,000

   3) transmission

   4) cloud

   5) compatible

   6) database

   7) app

   8) networks

   9) revolution

   10) worldwide

Unit 2　Energy and Technology

Listening

Researchers Look to Sodium to Make Better Batteries

1. **Listen to the recording and choose the best answer to each of the following questions.**

   1) C　　2) A　　3) C　　4) C　　5) B

2. **Listen to the recording and fill in the blanks with what you hear.**

   1) attempted

   2) power much-improved

   3) silvery metal

   4) build a new generation

   5) a done deal

   6) a work in progress

   7) testing new sodium batteries

   8) required to gather lithium

Smart Cities

1. **Listen to the recording and decide whether the following statements are true or false. Write "T" for true and "F" for false.**

   1) F　　2) T　　3) F　　4) F　　5) T

2. **Listen to the recording and answer the following questions.**

   1) In smart cities, buildings turn the lights off for you, self-driving cars find the nearest parking space, and even the rubbish bins know when they're full.

2) They will be able to use them to control the heating and locks, take part in video conferences, and receive education, healthcare and government services.

3) IBM, Siemens and Microsoft.

4) It gathers traffic data in Singapore which it uses to predict where traffic jams will occur.

5) Smart citizens.

Viewing

Top 10 Energy Efficiency Tips for Your Home

Video One

1. Watch a video clip about energy efficiency and decide whether the statements are true or false. Write "T" for true and "F" for false.

   1) T   2) T   3) F   4) T   5) F

2. Watch the video clip again and fill in the blanks with what you hear.

   1) heat the air and water

   2) renovation upgrade report

   3) losing heat

   4) air leakage

   5) the basement or foundation

   6) exposed floors and windows

Video Two

1. Watch a video clip and rank the energy efficiency measures in the order of importance.

   5–1–6–3–4–2–8–10–7–9

2. **Watch the video clip again and fill in the blanks with what you hear.**

1) small and big changes

2) Insulation and other upgrades

3) sweat the little things

4) more than 50%

5) green energy futures

6) a solar system

7) at the same cost

8) pay a carbon tax

Unit 3   Energy and Ethics

Listening

Providing Electricity to Poor Communities in Kenya

1. Listen to the recording and choose the best answer to each of the following questions.

1) B   2) A   3) A   4) B   5) C

2. Listening to recording and fill in the blanks with what you hear.

1) launched a campaign

2) availability of electric power

3) human waste

4) illegal connections

5) offer of financial support

6) challenges from the slums

7) hack our power systems

8) decided to disconnect

U.S. Withdraws from the Paris Agreement

1. Listen to the recording and decide whether the following statements are true or false. Write "T" for true and "F" for false.

1) F   2) T   3) T   4) F   5) F   6) F   7) T   8) T

2. Listen to the recording and answer the following questions.

1) He has promised to rejoin the climate agreement if he is elected.

2) Such an increase could raise sea levels, fuel powerful storms and worsen droughts and floods.

3) China, Japan, South Korea, the European Union (EU) and other countries.

4) He supports the call.

5) Government spokesman Paul Seibert said, "It's all the more important that Europe, the EU and Germany lead by example."

Viewing

The Ethics of Nuclear Energy

Video One

1. **Watch a video clip and tick the advantages of nuclear energy mentioned.**

   1) \   2) √   3) √   4) \

2. **Watch the video clip again and answer the following questions.**

   1) This is mostly because of the relatively small amount of waste nuclear power plants pump into the air.

   2) A ton of radioactive nuclear waste will usually get put in containers and stored far underground.

   3) According to a 2013 NASA study, an estimated 1.8 million deaths have been prevented since 1976 due to the use of nuclear energy.

   4) The reduction immediately lowers their risk of lung or heart diseases.

Video Two

1. **Watch a video clip and decide whether the statements are true or false. Write "T" for true and "F" for false.**

   1) T   2) F   3) T   4) T   5) F

2. Watch the video clip again and fill in the blanks with what
   you hear.

   1) 50

   2) inhabitants

   3) comparison

   4) atomic

   5) Treaty

   6) peaceful

   7) controversial

   8) byproducts

   9) standpoint

   10) reactor

Unit 4   Energy and Sustainability

Listening

How to Increase Your Renewable Energy Use?

1. Listen to the recording and decide whether the statements are true or false. Write "T" for true and "F" for false.

   1) F   2) T   3) F   4) T   5) T

2. Listen to the recording again and fill in the blanks with what you hear.

   1) creation; harnessing; implementation

   2) solar panels

   3) ethanol fuel; gasoline

   4) wind turbine

   5) importance; available

Making Flying Green and Sustainable

1. Listen to the recording and answer the following questions.

   1) With the rise of budget airlines, the number of people taking to the skies is increasing.

   2) Making flying green and sustainable.

   3) Biofuels can reduce the carbon footprint anywhere between 50%–80% when you compare them to fossil fuels.

   4) Using less fuel per passenger compared with aircraft of a similar size; better aircraft aerodynamics; changes to ways aircraft taxi on runways; the use of lighter materials.

2. **Listen to the recording again and fill in the blanks with what you hear.**

   1) norm; environment

   2) Climate change; the amount of fuel

   3) technology

   4) carbon neutral

Viewing

Innovative Companies: Sustainable Energy

Video One

1. **Watch a video clip and match the figures in the left column with the facts in the right column.**

   1) C   2) B   3) A   4) E   5) D

2. **Watch the video clip again and fill in the blanks with what you hear.**

   1) integrating

   2) productive

   3) capturing

   4) benefit

   5) synthesis

   6) methane

   7) filling

   8) released

   9) circular

   10) thin

Video Two

1. Watch a video clip and decide whether the statements are true or false. Write "T" for true and "F" for false.

1) T   2) F   3) T   4) F

2. Watch the video clip again and answer the following questions.

1) generation, transmission and utilization;

sectors, pathways and time horizons

2) supply and consumption;

any excess energy is saved

3) a transportation system or industrial system;

seasonal storage to the power system

Unit 5 Energy and Civilization

Listening

India Plugs into Low-Cost Solar Technology

1. Listen to the recording and decide whether the statements are true or false. Write "T" for true and "F" for false.

1) T   2) F   3) F   4) T   5) T   6) F   7) T   8) F

2. Listen to the recording again and fill in the blanks with what you hear.

1) solar panel; lamps; laptops

2) solar engineers; formal education

3) infinite capacity; incredible inbuilt skill

4) solar cooker; remote villages; the environment; firewood

5) self-confidence

Shenzhen's Silent Revolution

1. Listen to the recording and choose the best answer to each of the following questions.

1) B   2) D   3) B   4) D

2. Listen to the recording again and fill in the blanks with what you hear.

1) 16,000; 160,000; 440,000; halved

2) charging piles

3) charging facilities

4) a huge investment in electric transport

5) accelerating; emission-free; hybrid; all-electric

Viewing

What Happens When You Change Energy Culture?

Video One

1. Watch a video clip and decide whether the statements are true or false. Write "T" for true and "F" for false.

1) T   2) F   3) T   4) F   5) T

2. Watch the video clip again and answer the following questions.

1) Transport agencies are going to build more roads and more road infrastructure.

2) One is the global financial crisis; the other is uncertainty about whether people have a job.

3) They are our actions or activities, the physical things we have and how we think where the norms are.

4) In New Zealand people aspire to walk, cycle and use public transport more than they do.

5) They can make policies to support people to actually meet their own aspirations.

Video Two

1. Watch a video clip and list the four different visions of the future of the transport system.

1) There is a rapid increase in the number of people who are traveling by air.

2) The electric vehicles would become far more common.

3) Efficiency measures will drive far more efficient combustion engines.

4) Self-driving vehicles no longer need to regulate whether somebody can drive.

2. **Watch the video clip again and fill in the blanks with what you hear.**
   1) passionate
   2) trade-offs
   3) evidence
   4) outcome
   5) complicated
   6) norms
   7) collective
   8) externally
   9) mechanisms
   10) stability

Unit 6　Energy and Humanity

Listening

Solar Energy Makes the Difference in Africa

1. Listen to the recording and decide whether the statements are true or false. Write "T" for true and "F" for false.

1) T　　2) F　　3) T　　4) F　　5) F　　6) T

2. Listen to the recording again and fill in the blanks with what you hear.

1) inexpensive, reliable energy delivery

2) sunshine; makes all the difference

3) investments; electrical connections

4) access to clean and reliable electricity

5) these technological and financial innovations

Lack of Access to Energy

1. Listen to the recording and answer the following questions.

1) Burning carbon fuels produces large amounts of greenhouse gases which cause climate change and have harmful impacts on people's well-being and the environment.

2) Without electricity, women and girls have to spend hours fetching water, clinics cannot store vaccines for children, many schoolchildren cannot do homework at night, and people cannot run competitive businesses. Slow progress towards clean cooking solutions affects both human health and the environment.

3) Governments should invest in renewable energy resources, prioritize energy efficient practices, and adopt clean energy technologies and infrastructure.

4) You can bike, walk or take public transport to reduce carbon emissions.

2. Listen to the recording again and fill in the blanks with what you hear.

1) a stable electricity supply

2) 9 out of 10 people

3) spend hours fetching water; store vaccines; do homework; run competitive businesses

4) Energy services; powering healthcare facilities; supplying clean water

5) maintain and protect ecosystems; renewable sources

Viewing

Humans and Energy

Video One

1. Watch a video clip and put a tick before the energy which comes from the sun.

1) √   2) \   3) √   4) √   5) √   6) √

2. Watch the video clip again and fill in the blanks with what you hear.

1) 99%; ultimately

2) variety; metalwork

3) notable; nutritious

4) derive; technically

5) utilized; originally

6) steam engine

Video Two

1. Watch a video clip and decide whether the statements are true or false. Write "T" for true and "F" for false.

1) F   2) T   3) T   4) F   5) T

2. Watch the video clip again and fill in the blanks with what you hear.

1) illumination; efficient

2) revolutionary; century

3) Soviet Union; enthusiasm

4) radioactive; bombs

5) casualties; evacuate

Unit 7   Energy and Environment

Listening

Tons of Methane Gas Could Be Trapped Under Antarctica

1. Listen to the recording and complete each of the sentences with a choice from the box.

   1) F; E

   2) D

   3) C

   4) A

   5) B

2. Listen to the recording again and fill in the blanks with what you hear.

   1) ice-covered

   2) the organic material; sediment

   3) land; surrounding; hydrates

   4) freeing; deposits

   5) release; speed up

Solar Plant Raises Environment Concerns

1. Listen to the recording and decide whether the statements are true or false. Write "T" for true and "F" for false.

   1) F   2) F   3) F   4) T   5) T

2. Listen to the recording again and fill in the blanks with what you hear.

   1) Solar Electric Generating System

2) concentrations

3) molten salt

4) in the long run

5) colliding; burns; intense

Viewing

Energy and Environment

Video One

1. Watch a video clip and choose the best answer to each of the following questions.

   1) B   2) A   3) D   4) D   5) C

2. Watch the video clip again and fill in the blanks with what you hear.

   1) smog-free

   2) purifier

   3) removes

   4) jewelry

   5) creation

   6) provider

   7) global

   8) store

   9) vital

   10) releasing

Video Two

1. Watch a video clip and decide whether the statements are true or false. Write "T" for true and "F" for false.

   1) T   2) T   3) F   4) F   5) F

2. Watch the video clip again and answer the following questions.

1) In African villages, there is no electricity, no running water, no paved roads. These Folks are living in huts, and they have no heat and they have no electricity.

2) The children go off into the fields looking for animal dung and bring it back to the hut. Mom uses it to cook and to heat the hut.

3) They spend hours collecting the dung, which prevents them from going to school.

4) It is extraordinarily toxic and more people die in Africa of respiratory disease than any other single cause.

5) It is completely self-contained.

Unit 8　Energy and Future

Listening

Islay's Green Energy Projects for the Future

1. Listen to the recording and choose the best answer to each of the following questions.

1) B   2) A   3) A   4) D   5) C

2. Listen to the recording again and fill in the blanks with what you hear.

1) model

2) Harnessing

3) commercially

4) potential

5) limitless

6) sponsored

7) feasibility

8) distilleries

9) charged with

10) tapping into

Scottish Whisky Tested as Alternative to Fossil Fuels

1. Listen to the recording and tick the statements which are true about biobutanol.

1) √   2) \   3) \   4) \   5) √

2. Listen to the recording again and answer the following questions.

1) Scotland.

2) Whisky comes from grain, such as corn, rye or wheat.

3) First, the liquid is purified. It is heated until it becomes a gas. The gas is then cooled, and the resulting liquid collected.

4) Pot ale and barley.

5) Bioethanol is made from plants like corn or sugar cane.

Viewing

Hydrogen—the Fuel of the Future

Video One

1. Watch a video clip and choose the best answer to each of the following question.

   1) D   2) A   3) A   4) D

2. Watch the video clip again and fill in the blanks with what you hear.

   1) temperature

   2) massive

   3) emergency

   4) passengers

   5) network

   6) charger

   7) luxuries

   8) facilities

   9) logical

   10) station

能源人文英语视听教程

Video Two

1. **Watch a video clip and decide whether the statements are true or false. Write "T" for true and "F" for false.**

   1) T   2) F   3) T   4) F   5) T

2. **Watch the video clip again and fill in the blanks with what you hear.**

   1) isolated

   2) rural

   3) landscape

   4) mainland

   5) wave

   6) candidates

   7) generate

   8) create

   9) birthplace

   10) zero-carbon

Scripts

Unit 1　Energy and Daily Life

Listening

Fuels Used in Our Daily life

The world depends on a great deal of its energy in the form of fossil fuels. Examples of fuels include gasoline, coal and alcohol. Most of the fuels come from non-renewable sources; once used, they are gone forever. Each day, people bathe, cook, clean, do laundry and drive using various types of fuels. A quick review of different fuels reveals the important roles they play in daily life.

Important fuels used in everyday life include gasoline, coal, natural gas and diesel fuel.

Gasoline—Essential for Transportation

The most obvious fuel used in daily life runs cars, school buses and trucks. Gasoline and diesel are non-renewable fuels created from crude oil deposits in the ground or beneath the oceans. Lawnmowers and other maintenance equipment also run on gasoline. Construction sites power backhoes, dump trucks, cranes and other equipment with diesel.

Natural Gas—Heating and Cooking

Natural gas can power the heating systems, stove tops, water heaters and dryers in your home. Natural gas burns very cleanly and creates abundant energy when burning, according to the Natural Gas.org. This type of fuel is mostly comprised of methane but can contain other gases as well. Natural gas often occurs as underground pockets near oil deposits. Oil emits gases that rise to the higher levels of underground pockets of oil trapped within rock layers. Wells tap into these pockets to remove the natural gas for use in your home.

Coal—Electric Power

Many electrical plants burn coal as the primary fossil fuel for powering the electrical supply for homes across the country. According to the American Coal Foundation, coal-powered electricity fuels the electrical needs for more than half of all U.S. homes. Machines crumble the coal into small particles that get placed inside a furnace. The coal gets burned to heat water and create steam that fuels a turbine to create mechanical energy. This mechanical energy converts to electrical energy in a generator then gets transmitted through substations that deliver electricity to customers.

Alcohol—Gasoline Helper

Alcohol has played a major role as a fuel supply in recent decades. In particular, alcohol, or ethanol, made from corn is mixed with gasoline for much of the U.S. liquid fuel needs. Properly designed, cars and trucks can burn the gasoline-alcohol mixture without problems. By adding US-made alcohol to gasoline, the country's fuel suppliers reduce the need for imported crude oil.

Uranium—Carbon-Free Power

Although uranium isn't "burned" to make heat like coal or

natural gas is, it still counts as fuel as nuclear power plants consume it and extract energy from it. It is also like coal or other fuels in that it is non-renewable: When the supply is used up, it is gone for good. Unlike fossil fuels, uranium creates heat through radioactive decay, a process that, weight for weight, can yield as much as 1 million times the energy. The downsides of uranium include dangerous radioactivity and waste that remains radioactive for thousands of years.

Water

Water is often called the fuel of life, and for good reason. Our bodies consist of 60 percent to 75 percent water. We use water to bathe, wash clothes, cook and drink every day. This form of fuel also generates power for homes in areas near running streams and rivers. Dams block the water flow, creating built-up energy as the water accumulates. When the sluices release, the water flows toward a large turbine. The energy converts from mechanical to electrical energy and then is transmitted to a transformer to boost the electrical output. This renewable source of energy limits air pollution and provides about 7 percent of U.S. electrical power, according to the U.S. Geological Survey.

Solar Energy

We benefit from the sun's energy every day. It heats the Earth, provides heat, fuels the water cycle that produces weather and helps plants grow. Sunlight helps our bodies generate Vitamin D to absorb calcium. Solar energy dictates our daily life patterns of rest and activity.

How to Reduce Our Carbon Footprint

By Ian Morrison

The recent spate of pollution to engulf Beijing—and many other cities in China—should lead us to the conclusion as to what measures we can take as individuals who are faced with this growing problem on a daily basis to help mitigate it.

China's impressive record of economic growth since the turn of the century has improved the lifestyles of many of its population in ways which would have been beyond recognition just a matter of a few decades previously.

But this has also led to the development of lifestyles which consume amounts of energy that would have been unimaginable several decades ago, with the carbon footprint of the average Chinese urbanite increasing from the mere imprint of a baby's footstep to one which would resemble a lumbering giant crashing about before us.

So the next time we complain or justifiably feel concerned about the dangerous levels of air pollution, we should also look ourselves squarely in the face and consider what we should do to help bring our own carbon footprint down by a few sizes.

Before you make your next car journey, you ought to consider whether there is any other viable alternative, or the next time you reach to switch on one of the myriad of electrical appliances in your home, you ought to consider whether it is absolutely necessary for you to do so.

You may be nodding your head in agreement with my points, but like many people in this increasingly individualistic society, your own personal concerns will probably come first, with those of your fellow citizens relegated to taking a "moral stance" on such "important" issues.

So let's be frank, most of us won't take the required measures on our own behalf, an element of compulsion, or better still, encouragement, is often required.

One method which has been proposed in the United Kingdom is Personal Carbon Trading, with each citizen issued with a certain number of points every year. Those who indulge in excessive use of energy would be required to trade points with those whose carbon footprint remains below the average level. That's the stick.

While the carrot could be that those accounts remain in credit at the end of the year could be rewarded in some way, such as discounts for public transport, or even discounts for purchases in selected retailers—there are many ways in which this system could be used.

Such a system could be based on an electronic card, similar to those we use in many other aspects of our daily life, and even linked to other systems such as the chips contained in public hire bicycles, meaning that the use of emissions-free forms of transport—such as cycling—could actually be rewarded with additional points, offering a positive incentive to those who choose to adopt a greener lifestyle.

Viewing

Solar Energy Products You Should Buy

Video One

HELIOS Headphone

HELIOS is a hybrid Bluetooth headphone that captures solar power. Thanks to its photovoltaic panel, it can charge its battery with clean and free energy. For one hour of solar exposure, HELIOS delivers half an hour of music at full power. With a full battery, HELIOS can deliver around 15 hours of music. We thought HELIOS is designed in terms of ergonomics and simplicity. Besides, it is an expression of freedom and infinite autonomy. HELIOS combines green energy and high-quality sound. It is equipped with a mini USB port, so you can top off its battery by connecting it to a power source. If ever you're out of battery you can continue listening to music by plugging HELIOS to your device using a standard audio jack. You will never run out of music. Naturally, HELIOS allows you to answer calls hands-free. Nothing is more simple than connecting HELIOS to your smartphone: Turn on Bluetooth; connect to HELIOS; and listen freely to your music. Our team has worked hard to create a product which is reliable and consistent with our vision of the future. We wanted to create headphone that is truly nomadic, respectful of the environment, and energy-independent.

Kali PAK

We all need powers to function, survive and connect. But we always seem to run out of it, either because we choose to get out to nature and away from the grid, or in the case of a natural disaster the grid chooses to get away from us. The rise and number of natural disasters and the effect it has on communities, has created a real need for an autonomous energy source to provide lighting, communications, and electric energy for a wide range of devices. This is why we created this—the Kali PAK, a green energy generator, that fits dozens of off-the-grid energy needs. The Kali PAK is completely sustainable and autonomous. It can rely solely on its patented solar panel with a unique angle dial and comes with a rain cover, so you can keep the power going in case of rain. The Kali PAK app connects through the built-in Bluetooth and lets you follow the PAK's energy levels and effectiveness, connected devices' consumption levels, and more. The 39-amp battery is powerful enough to charge your iPhone up to 120 times, an iPad up to 40 times, and in our case, powerful enough to provide energy for our entire set for a whole day including a fridge till we ran out of beer. When designing the Kali PAK, our team of industrial designers, mechanical and electrical engineers, battery experts, and disaster experts focused on developing a seamless user experience functionality in different grid-independent scenarios. Your backing will allow us to bring the Kali PAK to market and provide a valid solution for all of our off-the-grid energy needs for emergency situations, or just you'll be able to keep the fun going. In addition, your backing will allow us to donate Kali PAK to African communities who lack energy infrastructure and make their life a little easier. So may the power be with you whenever and wherever you need it.

Video Two

ODO

Don't you feel guilty about wasting water? Everyone likes a beautiful garden or a terrace full of colorful plants. But gardens and plants need fresh water and reserves are being depleted globally faster than you think.

Lorenzo Torracca: The planet is suffering and we should all be on a mission to reduce the impact that our human activities make on it. We have modern appliances, we use clean energy. However, we still irrigate like we did hundreds of years ago. This inspires me to look for a smart system that through the use of the latest technology would be able to reduce the impact of irrigation on potable water reserves. So, I began the development of ODO. ODO is the most advanced, efficient, sustainable, smart, domestic irrigation solution system.

ODO is a complete and flexible solution for the irrigation of your garden or terrace. ODO is simple and easy to install. It knows the water needs of your plants and automatically adjusts the irrigation profile to weather forecasts and to soil conditions. This flexibility improves efficiency and water saving.

Lorenzo Torracca: It allows you to save more than 7,000 liters of potable water each year, keeping the highest standards of health and beauty for your plants. It makes use of clean energy from the sun. It is completely self-sufficient and sustainable.

Our technology for data transmission is the core of the product. It works at a low frequency that allows a wider range key and low battery consumption. ODO is connected to the cloud and constantly monitors the weather forecasts from the web. It also compares these data for a double-check with its embedded weather station. It's plug-and-play and compatible with every existing irrigation system. ODO is completely automatic and easy to set up: You can manage your entire system by the palm of your hand and control the status of your garden wherever you are. ODO takes care of your plants for you and you can easily control the entire process. It has an extensive plants database to control your cultivation through its sensors. You can always know the health status of your plants. The ODO app is the perfect tool to monitor and manage your system wherever you are. You just need an Internet connection. ODO is a social movement for water sustainability too: You can share your experiences and your successes on popular social networks or on its dedicated community. Now you can open your faucets, being sure not to waste water. Now you are part of a revolution.

Lorenzo Torracca: We are offering an affordable price to make this revolution possible. We want to spread ODO all over the world, but we need your support. The more ODO is spread worldwide, the more affordable will be our product and we'll make a larger revolutionary impact on our planet. This isn't just about smart gardening, this is a social movement for water sustainability.

Unit 2   Energy and Technology

Listening

Researchers Look to Sodium to Make Better Batteries

Scientists have long attempted to find materials to make batteries that are more powerful, but cost less to build.

In the United States, researchers are experimenting with sodium to see whether it can power much-improved batteries in the future.

Sodium is a soft, silvery metal. It is plentiful and found in seawater.

The most common battery used today is made of lithium ion. These batteries power everything from smartphones to computers to electric vehicles.

Researchers from the University of California, San Diego, are attempting to build a new generation of batteries powered by sodium instead of lithium. The U.S. National Science Foundation is providing financial support for the experiments.

Shirley Meng is a member of the research team.

"At a society level, I think people really think that a battery is a done deal—like it's an old object."

But Meng says the process of developing better batteries is still a work in progress. In fact, she says the energy density of batteries in use today "can still be doubled or tripled".

The California researchers are studying lithium ion batteries, but in the next few years plan to begin testing new sodium batteries. Team member Hayley Hirsh says she looks forward to working more with sodium development in the future.

"We want to use sodium instead of lithium because it has different properties. And also, sodium is much more abundant."

Lithium is costly and not easy to collect because it is widely spread across many parts of the world. Large amounts of water and energy are also required to gather lithium.

But sodium is found in the world's oceans, with a seemingly limitless supply. This would lead to much lower costs to produce sodium ion for batteries.

Hirsh says she is examining different ways to make batteries that last longer and can store more power.

"Right now it's just in the lab and we're working on figuring out how to make it hold more energy and last longer so that it can be used in your phone, in your car or even to store energy for solar, for wind."

Finding better ways to store more energy at a lower price has been one of the major barriers to developing more powerful batteries.

Today it is not really cost-effective for power companies to use batteries. This is because it would cost hundreds of dollars per kilowatt hour to operate.

However, using sodium ion batteries could bring that cost way down. The researchers say it could then make economic sense for people to have storage containers at home to save energy produced by the sun or wind.

"They have solar on the roof. They could store the electrons during the day and use them at night," Meng said.

Smart Cities

Would you like to live in a city where buildings turn the lights off for you, where self-driving cars find the nearest parking space, and where even the rubbish bins know when they're full?

Although it might sound like science-fiction, living in a "smart" city like this could happen sooner than you think.

Towns we have lived in for centuries are being upgraded, while completely new cities are being built.

One such place is Songdo in South Korea. Every home there will have a built-in "telepresence" system, allowing users to control the heating and locks, take part in video conferences, and receive education, healthcare and government services.

Around the city, escalators will only move when someone is on them, and offices and schools will all be connected to the network.

The digital mastermind behind Songdo is the company Cisco. Indeed, technology firms around the world such as IBM, Siemens and Microsoft are already selling software to solve a range of city problems, from water leaks and air pollution to traffic congestion.

For example, IBM is gathering traffic data in Singapore which it uses to predict where traffic jams will occur—an hour before they happen.

So what is all this smart technology for? Many hope all this connectedness will make cities greener, more sustainable, and more efficient.

And with 75% of the world's population predicted to live in cities by 2050, the transport system and emergency services will need to modernize to cope with all the new arrivals.

But making cities smarter is only part of the solution, according to Dan Hill, chief executive of research firm Fabrica.

He said: "We don't make cities to be efficient, we make cities for culture, commerce, community—all of which are very inefficient."

In the rush in order to make cities perform better, we could be missing their greatest asset.

"It is going to be smart citizens that make smart cities," he said.

Viewing

Top 10 Energy Efficiency Tips for Your Home

Video One

Jeff Paton: This is your EnerGuide Homeowner Information Sheet. We can see here that your energy rating and gigajoules per year is 236, that's a significant amount of energy. But it's not uncommon for this area of construction in the 1950s. This is a great tool that we can use to understand how the energy is used in your home. This particular home is using 80% of its energy on space heating, another 10% on water heating, and the remaining 10% on lights appliances and other electricals. Typical homes, around 65% of their energy used on space heating, so you're a little higher than that. But again, for the age of construction, it's pretty typical.

Reporter: That's Jeff Paton of Sunridge Residential walking Brian Finley through EnerGuide for Home's Assessment Report. Finley's a real estate agent from Edmonton Alberta.

Brian Finley: I really wanted to know more about it, and sometimes the best way to know about something is to actually do it. In this case, I went out and actually scheduled the audit, so they'd come in. I could understand what it was so that I could see how it impacted my own home.

Reporter: Our goal at green energy futures was to follow Jeff as he assessed Brian's home. Then, build a list of the top ten energy efficiency actions you can take in your home. As Jeff explained, the EnerGuide Assessment told Bryan that his 1956 home uses a whopping 236 gigajoules of natural gas and electricity per year. If Brian's home were built to today's standards, it would consume less than half of that.

Brian Finley: While a lot of people said, you're going to have your 1956 house looked at for energy efficiency. I mean, the number is what the number is, so you might as well, figure out what it is and then figure out what you want to do to make it better.

Reporter: Brian's 1956 home uses 90% of its energy to heat the air and water in the home. Not surprisingly, the renovation upgrade report focused on things that will save energy used to heat the home.

Jeff Paton: This is another useful tool that we can look at where the home is losing heat. In this instance, this home is losing 29% of its heat through air leakage, another 25% of it through the basement or foundation, and the rest is made up to attic, main walls, exposed floors and windows.

Reporter: Remember, every home is unique. To build our top 10 list of energy efficiency measures, we followed Jeff around Brian's house to pick up some tips, and then we added some measures that aren't covered in an EnerGuide Home Assessment.

Video Two

Home Energy Assessment

First on our list, do a Home Energy Assessment. This will tell you where the big potential energy savings are. Then, you can make tough informed choices about what to do.

Brian Finley: We know we are going to do a new roof, so the difference now won't be that we would do the new roof. It would be when we do the new roof; it would be what are the things can we do; it would be the perfect time, obviously, to add insulation to the attic.

Insulation

Insulation is No. 2 on our top 10 list. Most of Brian's proposed measures involve adding insulation. And it almost always offers a good return on investment.

Seal the Envelope

We've already mentioned No. 3, which is to seal up the building envelope, reduce air leakage and save energy. Air seal was near the top of the proposed measures for Brian's home.

Windows

Brian Finley: We did our windows probably two years ago on the second floor, so we redid all the windows up there. We have yet to do the windows, and some of the areas down here on the first floor. You can do it in phases. It's like everything else, you can't do it all at one time.

Furnace

Next up, No. 5, the mighty furnace and something the EnerGuide Assessment doesn't touch on.

Smart Thermostat

No. 6, the smart thermostat. Brian's furnace is already mid-efficiency, but your 20-year-old furnace is probably about 77% efficient. You can gain 20% of efficiency from improving the furnace, and between 10% and 30% from a smart thermostat. Those are two measures that are the easiest way to save up to 50% on home heating.

Appliances

Jeff Paton: Well, we've got a number of appliances in the home that are going to account for a large percentage of the energy used in the electrical consumption. Almost all the appliances we have in our homes today have improved over the last 20 years by 100%, washers, dishwashers, fridges especially. There are a couple of things that haven't changed.

Reporter: Dryers and stoves haven't improved much, but the new heat pump dryers and induction stoves could change all that, as they become more common in North America.

Water Heater

Reporter: After that is No. 8 on our list—the water heater that uses 19% of the energy in your home.

Jeff Paton: Naturally aspirated water heater like this is going to be around 60% to 65% efficient, and it's very easy to go to a tankless style water heater that are up around the neighborhood of 95% to 97% efficient.

Reporter: There are also 90% efficient power vent water heaters and 250% efficient heat pump water heaters.

Lights

Jeff Paton: Energy-efficient lighting is going to be one of the easiest things that we can do, and it's going to have a very short payback period under a year. LED bulbs are 75% to 80% more efficient than incandescent bulbs.

Reporter: Lights and appliances are big users of electricity, replace old incandescent bulbs with LEDs and be sure to choose Energy Star appliances.

Phantom Power

Reporter: Finally, No. 10 on our list is something called phantom power. Countless electronic devices—TVs, computers, printers, game systems and device chargers—draw power when they're not even being used. Exercise these demons by unplugging devices or by using a smart power bar. Leaving your printer on is like leaving a light on

all day, 24/7, 365 days a year.

So there you have it, our top 10 list of energy efficiency tips for your home. Remember, every home is different. But by checking these things in your home, you will find small and big changes that can make a big difference in your home. Brian Finley now knows where his home can be improved. Insulation and other upgrades can make a big difference. But when it comes to energy efficiency, you just might want to sweat the little things as well. We've seen homes where owners have reduced their energy consumption by more than 50% by simple measures, such as changing lights to LEDs, adding smart thermostats, unplugging the monster beer fridge, and taming their energy phantoms.

As we've learned on green energy futures, it's always a good idea to make your home energy-efficient first. But having done that, you just might want to add a solar system to your home and produce your own energy. You'll save money on electricity over the life of the system and you'll be producing your own clean energy at the same cost for the next 25 years. As a bonus, you won't pay a carbon tax.

Unit 3   Energy and Ethics

Listening

Providing Electricity to Poor Communities in Kenya

In 2013, President Barack Obama launched a campaign called the Power Africa Initiative. Its goal is to increase the availability of electric power in African countries south of the Sahara Desert. Millions of people there are unable to depend on reliable power supplies.

The American-supported program is providing money for a number of projects, including one that creates electricity from human waste.

"Mukuru Kwa Njenga" is the name of a community close to Nairobi where about 100,000 people live. Many of them are poor. Until recently, most did not have electricity. Those who did had illegal connections to power lines.

Amos Nguru had the idea for Afrisol Energy, a project that produces electricity from human waste. Two years ago, Mr. Nguru received an offer of financial support from the Power Africa Initiative. The money came from General Electric and the United States African Development Foundation.

The project now produces 15 kilowatts of electricity. That is enough to power a nearby school and serve the local neighborhood. Mr. Nguru says his project meets the needs of the community.

Deborah Mwandagina is deputy head teacher of the local primary school. She says in the past that there were too many illegal connections to her school's power supply. She says this resulted in higher costs for the school, so it decided to stop using electricity.

She says, "We have been having challenges from the slums and time and again they hack our power systems and they connect illegally to their homes. We have been having such kind of challenges so we just decided we cannot live with electricity. It can't do us any good, so we decided to disconnect."

But now the school is once again using electricity from Afrisol Energy. Doreen Kemunto is a student at the school. Her mother Beatrice Onchan'ga says darkness no longer limits the time when Doreen can study.

She says, "Before the electricity was there, our children could not be able to learn since they could not come to school that early and leave that late because the school was very dark, and they could not be able to learn."

The World Bank says only 23% of Kenyans have access to electricity.

U.S. Withdraws from the Paris Agreement

The United States officially withdrew from the Paris Agreement to fight climate change on Wednesday. Leaders from around the world approved the agreement in 2015 at a conference in the French capital.

For more than two years, American President Donald Trump talked about withdrawing from the treaty. Last year, the Trump administration announced the U.S. decision to withdraw. However, the results of the presidential election could decide for how long. Trump's main opponent in the vote, former Vice President Joe Biden, has promised to rejoin the climate agreement if he is elected.

More than 180 countries remain committed to the 2015 Paris Accord. The agreement aims to limit the increase in average temperatures worldwide to "well below" 2 degrees Celsius, and ideally no more than 1.5 degrees Celsius. Those increases are compared to temperature levels before Europe's Industrial Revolution.

Scientists say that any temperature increase greater than 2 degrees Celsius could have a disastrous effect on large parts of the world. Such an increase, they say, could raise sea levels, fuel powerful storms and worsen droughts and floods.

The Paris Accord requires countries to set their own targets for cutting production of carbon dioxide and other gases linked to rising temperatures. The only legal requirement is that national governments must truthfully report on their efforts.

The United States is the world's second biggest producer of heat-trapping gases, after China. In recent weeks, China, Japan and South Korea have joined the European Union (EU) and other countries in setting national targets to stop pumping more greenhouse gases into the atmosphere.

Biden, the Democratic Party's candidate for president, has said he supports calls for the United States to return to the Paris Accord.

On Wednesday, Germany's government said it was "highly regrettable" that the United States had left the Accord. "It's all the more important that Europe, the EU and Germany lead by example," government spokesman Paul Seibert said.

While the Trump administration has rejected federal measures to cut greenhouse gases, Seibert noted that U.S. states, cities and businesses have pushed ahead with their own efforts.

Viewing

The Ethics of Nuclear Energy

Video One

What is nuclear energy? Well, for starters, it's not a new technology. It's been around since before the 1970s, you know, back when everybody looked like this.

Nuclear energy is possibly one of humanity's most untapped potentials in terms of technology. Yet, also one of the most lethal. Today, I'm going to talk to you about the pros and cons of nuclear energy. I believe everyone should be informed about technology such as nuclear energy as it could make or break society as we know it.

Pros

Start off on a positive note with the pros of nuclear energy. Don't worry, we'll get to the heavy stuff soon enough. First of all, nuclear energy is really good for the environment. Okay, not "really good". But compared to what fossil fuels are doing to our planet, nuclear energy is looking real good right now. This is mostly because of the relatively small amount of waste nuclear power plants pump into the air. Of course, there is a ton of radioactive nuclear waste that nuclear power plants produce, they usually get put in containers and stored far underground. While that might not seem like a particularly bright idea, it's better than pumping it back into the atmosphere like most fossil fuel plants do.

It also has saved many lives, kind of. According to a 2013 NASA study, an estimated 1.8 million deaths have been prevented since 1976, due to the use of nuclear energy. This is mostly because any reduction to the amount of fossil fuels in the air immediately lowers a person's risk to lung or heart diseases due to pollution. It also causes the least deaths per unit of energy per year in its field with fossil fuels as the leading energy-related cause of death. So, sure, it might not damage people as badly as our current energy options, but there are tons of people who are very against nuclear energy. One of their main arguing points is that most nuclear power innovations have been put on hold since the mid to late 1900s. This means, most of the technology is outdated and certainly not fit to function positively in this day and age. Plus, many say that this lull in the exploration and innovation when it comes to nuclear energy proves that the world does not want nuclear energy. But supporters will say that while outdated technology is not as effective if exploration of the topic continues, it's highly likely that they would be able to fix a lot of the problems that nuclear reactors currently have. Their most reasonable proposal is the thorium reactor. Currently, most nuclear power plants function with the help of a plutonium reactor. Plutonium is also one of the key aspects of current nuclear energy. And while this does its job, it is one of the things that makes both nuclear weapons and nuclear waste so dangerous. In fact, just one milligram of plutonium could kill you.

By using a thorium reactor, the risk would be avoided as thorium is sufficiently less dangerous. The waste products would still hypothetically have radioactivity but it would dissipate after a couple hundred years. Though this may not sound good, that's because you didn't know the current plutonium-infused nuclear waste takes tens of thousands of years to dissipate in terms of radioactivity. Thorium would be a wonderful alternative. Unfortunately, due to most countries' hesitancy to meddle with anything concerning nuclear energy, no one has really been able to test this. However, this hesitancy is completely justified. How? Well, let me tell you some of the cons—the bad things about nuclear energy.

Video Two

Cons

First, let me say these, Chalk River, Kyshtym, Sellafield, Lucens, Three Mile Island, Chernobyl, Fukushima, sound like a bunch of random gibberish. Well, it's not. What I just said was a list of the seven major nuclear disasters that have happened around the world in the past 50 years. Though three of these at ordeals were fairly self-contained, the other four actually render entire parts of countries unfit for human inhabitants. The most recent nuclear emergency was Fukushima, more formally known as the Fukushima Daiichi disaster. It happened on March 16, 2011. There was an energy accident at the Fukushima nuclear power plant. No one died from the major radiation leak, though 37 people were injured and 2 workers were hospitalized due to radiation burns. The outcome of this accident was a rise in thyroid cancer among the people living near the power plant of Okuma, Japan, as well as a general redressing of the issue of nuclear energy, to which most of the world show general descent on any encouragement of nuclear energy.

Another issue with nuclear energy is nuclear power plant waste and pollution. Like I mentioned earlier, it is healthier for the environment but only in comparison to fossil fuels. The amount of waste is not the issue but it is the dangerous properties of the side waste that make it such a concern. Always, nuclear power plants is radioactive and contains plutonium, which like I said before, is a large threat to us as humans. For one, just a milligram of plutonium could kill a human being and a kilogram is all you need to make an atomic bomb. Atomic bombs, of course, and general weaponization is the most concerning part of nuclear energy. Though there are a total of 439 nuclear reactors in 31 countries, most of them were built under the presumption that they were going to be used for energy production and not for weapons. Sadly, in just under 40 years, five countries have developed their weapons using nuclear energy. So, it is difficult to tell when a country intends to create nuclear weapons and when it is intending to use nuclear power plants for what they were originally intended for—to power things.

You'll be glad to know that America is one of 190 parties to have the Nuclear Non-Proliferation Act, which is part of the Treaty on the Non-Proliferation of Nuclear Weapons that was created in 1968. The only parties that did not sign this treaty include India, Israel, North Korea, Pakistan, and South Sudan. The Nuclear Non-Proliferation Act aims for, and I quote, "more effective international controls over the transfer and use of nuclear materials equipment and nuclear technology for peaceful purposes to prevent proliferation". Proliferation, by the way, means a rapid increase in numbers. It can be exemplified by the nuclear energy craze of the 1980s during which a sum total of 218 nuclear energy reactors were created. Luckily, that time is past and we are in a present that attempts to revolve around the fair discussion of any issue.

In the context of this, nuclear energy can be an extremely controversial talking point. If you were to look at it morally and ethically, it doesn't seem to be hurting many people yet and could provide better electrical opportunities for many places in the world. But one of the byproducts of nuclear energy is the creation of very powerful nuclear weapons. And for this reason and this alone, it is hard to agree with nuclear energy from a moral standpoint. This is also the stance that I take as an individual, concerning nuclear energy. Though I am for the benefits of nuclear energy, I think that until we know for certain that we can control the effects of a nuclear reactor and what people use them for, it is not a good idea to encourage the use of nuclear energy or nuclear power plants.

We should take care of our planet and enter into an honest

dialogue as a whole about what is best for us and the world we live in. This is the same kind of dialogue I hope to pursue with this video. So, decide for yourself. Should we find ways to improve the world by going on the path less traveled, the path of improvement through nuclear energy? Or, should we set aside nuclear energy and focus on improving our world in more concrete ways, like reducing our carbon footprint? The decision is up to you. What will you choose?

Unit 4   Energy and Sustainability

Listening

How to Increase Your Renewable Energy Use?

Renewable energy involves the creation, harnessing and implementation of sustainable resources. This means that the source from which the energy is cultivated can be replenished and reused over and over again. Fossil fuels, long the world's standard energy sources, are quickly running out and are damaging to the environment. Renewable energy is the solution to these problems because it is not only unlimited in supply, but also environmentally friendly. Every citizen must strive to increase renewable energy use so that these technologies can continue to grow.

One way you can increase your renewable energy use is by having solar panels installed on your home or office building. These small, dark panels contain wires and receptors that capture and use the sun's energy to power homes and other buildings, without the use of limited resources. Solar energy panels can be pricey, but as demand continues to skyrocket, the prices are steadily declining.

Putting ethanol fuel in your gas tank instead of gasoline is another way to increase your renewable energy use. This liquid fuel source is comprised of corn alcohol and gasoline and helps lower the carbon emissions that are produced as you drive. Although it is partially comprised of a traditional fossil fuel source, it may soon be possible to find fuels made only from corn and other renewable sources. Not every car can run on ethanol, so you should be sure yours is able to do so, in order to avoid damaging your automobile.

Wind energy is another growing renewable energy source. You can take advantage of wind power by purchasing a small wind turbine and installing it at your home. There may be companies in some areas who will come and install the kit for you. Although wind energy is fully renewable and sustainable, it is also limited. This is mostly due to the unpredictability of wind, as some days are much windier than others. Even so, wind power is a great supplemental power option and will help to increase your renewable energy use.

Renewable energy sources will continue to grow in importance across the globe, although it may take some time before options are available in all areas. You can help bring more of these sustainable energy sources to your city, as well as encourage companies to implement them in the making of their products, by petitioning to the local government. Write letters to legislators stating your concerns and desires for cleaner energy. You can also take action by being a discerning customer regarding which products you buy. If more people lobby to purchase from businesses that support environmental causes, more companies will begin using them in order to maintain sales.

Making Flying Green and Sustainable

Flying has shrunk the world! It's now possible to travel around the globe quickly and easily. Jumping on a plane and jetting off on holiday or a business trip is the norm for many of us, and with the rise of budget airlines, the number of people taking to the skies is increasing. But while air travel is costing us less, the cost to the environment is going up.

Climate change is something we're all aware of now, and aviation companies know that some of the blame for this is being pointed at them. Last year, commercial airlines were forecasted to use about 97 billion gallons of jet fuel. But while we could think twice about taking a flight in the first place, particularly a short-haul trip that could be made by train, technology might be the answer to reducing emissions and minimizing the environmental damage.

Recent developments have focused on reducing the amount of fuel airliners burn. Making flying green and sustainable was the hot topic at the recent Dubai Air Show. There was talk of advances in engine technology, making them more efficient, and possibly using biofuel to power aircraft. Alejandro Rios Galvan, a bioenergy expert and professor at Khalifa University in Abu Dhabi, told the BBC that "these have the capacity to reduce the carbon footprint anywhere between 50%–80% when you compare them to fossil fuels." And Phil Curnock, chief engineer of the civil future programme at Rolls-Royce, also suggested that electric hybrid engines could play a part for smaller aircraft, covering shorter distances. He says, "It offers the possibility of a carbon-neutral flight for a limited range."

Being carbon neutral is the ultimate goal for the aviation industry, and one British airline, EasyJet, has recently said it would become the world's first major net zero carbon airline by offsetting carbon emissions.

Of course, aircraft manufacturers are constantly looking at ways to make their planes more fuel-efficient. Boeing's Dreamliner, for example, is already in operation and uses 25 percent less fuel per passenger compared with aircraft of a similar size. Other improvements include better aircraft aerodynamics, changes to ways aircraft taxi on runways, and the use of lighter materials.

But if we can't kick the flying habit, it seems these are the developments we need to make in order to ensure air travel is as green as possible. But aviation experts agree it's going to take time. We're left with short-term action such as taxing flights, regulation or protest—or being grounded and not flying at all.

Viewing

Innovative Companies: Sustainable Energy

Video One

Host & Hostess: As our energy horizons expand, so do our energy systems. Gone are the days when we would focus on one energy source at a time. Today we're looking at innovative multi-energy solutions designed to multitask and make way for a low-carbon future.

Host & Hostess: We see how indoor air pollution could fuel the future. We drive away from carbon emissions in a clean car and we meet scientists riding the wave of the renewables boom. And I traveled to Kenya to meet Benson Aery, an Africa lead for energy access at the World Resources Institute, to get his perspective on the benefits and challenges of multi-energy solutions. But first, some facts and figures.

Facts and Figures

It is predicted that 21% of global electricity production will come from variable renewables by 2040, up from 7% in 2018 supported by around 5.3 trillion dollars of investment. Experts expect that approximately 39% of electricity production in the EU will be generated from variable renewables by 2014. Battery energy storage systems look set to play a pivotal role in the future of renewables. It is estimated that the market for them will more than double, from 6.1 billion dollars in 2018 to 13.1 billion dollars by 2023.

Host & Hostess: Multi-energy solutions mean that almost anything can be thrown into the energy mix. In Finland, there's a start-up using solar- and wind-generated electricity, carbon sequestration, and water to make a multitude of sustainable products.

Mindful of the high levels of CO_2 in indoor air, the company plans to use its technology to clean what we breathe in our offices by integrating its technology in buildings. It says that this will improve people's well-being and make them more productive all while producing valuable clean resources with the CO_2 pollution.

Petri Laakso: First, we have here city air. You always have ventilation in the building and you push it through the ventilation unit through our carbon capturing unit, and then you are able to get lower CO_2 air indoors, which will be the benefit of people's well-being. Then we have the electrolyzer. We break down water into hydrogen and oxygen and then we have a synthesis unit which can do the fuel part or hydrocarbons. And if this building would be connected to a gas grid, you could provide synthetic methane which you can pump into the gas grid. So the gas grid could be added as energy storage, or we could have a car filling station that you can fill up the car tank.

The main aim is to use the hydrocarbons to fuel vehicles where they can be mixed with fossil fuels, and any CO_2 emissions released from the engine can then be pumped back into the solar tier process, closing the circular loop. The product can also be refined to produce a cleaner gas to heat and power homes. Scientists often wish they could pluck solutions for a sustainable future out of thin air. The solar tier team might just come close.

Video Two

After the break, we take to the wheel in a car that could steer the transport sector in a new direction. A key part of multi energy solutions is clean fuel, and in Wales, there's a company taking the concept of zero emissions up again. Riversimple, based in the countryside of the preserved County of Powys, is on a mission to eliminate the environmental impact of personal transport. A number of automakers have developed hydrogen-powered vehicles already, but this company says its model has the lowest carbon emissions of any vehicle. In fact, the only emissions it produces is water. Plus it's fuel-efficient. It can travel up to 300 miles on 1.5 kilograms of hydrogen—that's the equivalent of nearly 250 miles per gallon.

You thought you knew? Think again. Myth—multi-energy plants need to exist at the source, otherwise siting and transmission costs can be prohibitively expensive. Fact—whilst finding suitable locations for multi-energy plants can be difficult, and negotiations, permits and community relations can increase costs or delay projects. It has been proven possible to overcome these challenges. Firstly, let's define multi-energy systems. It's a site that coordinates the generation, transmission and utilization of energy, especially renewable energy across different energy sectors, pathways and time horizons. A prime location is one which offers ample sun, wind and land, all more commonly found outside of population centers. As a result, increased transmission infrastructure is necessary to connect supply and demand, which can affect system costs.

After the break, we see how multi-energy solutions cope when renewables are at an all-time high. When it comes to clean energy, it may just be possible to have too much of a good thing. In Beijing, there's a scientific duo equipping multiple energy systems with ways to cope with record volumes of renewables. China is known as a leader in renewable energy and technological innovation, but this influx of clean energy comes hand in hand with challenges, such as large-scale curtailment due to lack of power system flexibility. One effective solution is integrating multi-energy systems as Professor Chongqing Kang, Dean of the Department of Electrical Engineering at Tsinghua University, and Dr. Ning Zhang have found out.

Dr. Ning Zhang: Fundamentally, the driven factor of the multi-energy solution is that the technical characteristics of different energy systems are quite different. For example, electricity is easy to transmit and easy to use. But it needs to be real-time balanced. But batteries are still expensive. However, gas and heating system is significantly cheaper than storing electricity, so we can use it to indirectly store electricity so that we can accommodate more renewable energy.

Multiple energy systems increase the efficiency and flexibility of both energy supply and consumption, so they can be used when and where they're needed, and any excess energy is saved as opposed to getting tailed and wasted.

Dr. Ning Zhang: Jilin Province is a very cold province with six months of centralized heating. It is also very rich in wind power. the multi-energy solution there is to use electric boiler and heating storage, to use the wind power that might have been curtailed at night to produce heat. For one square meter of heating space, it can utilize 160 kilowatt-hours of wind power. That means 51 kilograms of coal. The overall project can save 8,000 tons of coal per year.

The team here has found that it's impossible to take a one-size-fits-all approach to multi-energy systems, as different regions of China have different resources and energy demands, and they all face their own unique challenges when attempting to fully accommodate renewable energy.

Dr. Ning Zhang: Southwestern China is rich for hydropower. One thing that we should know about hydropower is that it has a strong seasonal pattern. That means in a flooded season there is too much water to produce electricity that the load is not enough. Part of the gas system can use the energy surplus to produce a lot of gas using the hydropower. The gas can be used in a transportation system or industrial system or even to produce electricity in the dry season so that it can provide a seasonal storage to the power system.

The research of Dr. Zhang and Professor Kang makes the multi-purpose uses of multi-energy systems clear.

Prof. Chongqing Kang: According to the research of our five-year projects, by 2030 more than 35% of the energy demand will be supplied by the renewables. The number will reach 60% to 70% by the year 2050.

With such detailed research and fine-tuning, multi-energy systems could ease the high penetration of renewables, all while ensuring our energy systems are flexible and energy-efficient.

Host & Hostess: So, it looks like multi-energy solutions are truly multi-purpose. They help integrate more renewables into the system, clean up the transport sector, the air we breathe, and how we power our homes, all the while enhancing energy efficiency and flexibility.

Unit 5   Energy and Civilization

Listening

India Plugs into Low-Cost Solar Technology

India seems to excel at making things smaller and cheaper. The $2,500 car and the $35 computer are just two of the country's latest innovations. Now, India increasingly is focused on low-cost solar technology. The front lines of that effort are seen in a tiny village in the Indian state of Rajasthan called Tiloniya.

In this sunlit workshop, Tenzing Chonzom solders parts onto a device that regulates electrical currents. It will eventually be connected to a solar panel, allowing it to power everything from lamps to laptops.

Make Low-Cost Solar Panels

Chonzom says she was chosen by her community to come here to learn about solar technology. She says she will take the knowledge back to the villages where she lives. She says many people in her region, in the Himalayan foothills, still do not have access to electricity.

Chonzom is 50 years old, and one of two dozen people being trained here as solar engineers. Most have had no formal education. It is all part of a program to help India's rural poor by teaching them to make and install low-cost solar panels. Then they teach others to do the same. It is called Barefoot College, and so far it has trained thousands. Sanjit Bunker Roy started the program 25 years ago.

"You have to see how you can demystify the technology and bring it down to the community level so that they can manage, control and own the technology," said Roy.

Roy is among *Time* magazine's top 100 most influential people for 2010. He says grassroots solar technology is crucial for India. Nearly half the country's rural population—more than 300 million people—has either no electricity or just a few hours of it a day.

That limits how much people can do in a day, whether its homework or handicrafts.

Tapping Local Ingenuity

To help, Roy says he did what World Bank and UN aid projects often fail to do—that is, tap into the local ingenuity he sees every day.

"You'll find it everywhere in India, this infinite capacity to be able to improvise and fix things without having gone through any formal education," he added. "They have this incredible inbuilt skill that we haven't been able to define, appreciate or respect yet."

Women Built Solar Cooker

As if to illustrate their inbuilt skill, Roy points out a solar cooker that some of the women at Barefoot College helped design and build. At its heart is a "solar tracker" made of old bicycle sprockets, springs and rocks. It allows a parabolic mirror—also homemade and about the size of a satellite TV dish—to follow the arc of the sun, focusing its rays into an aluminum stove. All the meals at the college are cooked on it.

One of its main designers is Sita Devi, a 30-year-old mother of two with only a third-grade education.

She says she wanted to make a solar cooker with materials that are easily available, even in remote villages. She says the cooker saves time—and the environment—by reducing the need for women to wander outside the village in search of firewood for cooking.

In the sun, Devi and other women of the Barefoot Brigade extol the benefits of their solar cookers and lamps.

Roy says his program merely provides the space for women to develop self-confidence. He says that is what drives the Barefoot Brigade's success in bringing power to more than 450 rural villages.

Shenzhen's Silent Revolution

You have to keep your eyes peeled for the bus at the station in Shenzhen's Futian central business district these days. The diesel behemoths that once signaled their arrival with a piercing hiss, a rattle of engine and a plume of fumes are no more, replaced with the world's first and largest 100% electric bus fleet.

Shenzhen now has 16,000 electric buses in total and is noticeably quieter for it. "We find that the buses are so quiet that people might not hear them coming," says Joseph Ma, deputy general manager at Shenzhen Bus Group, the largest of the three main bus companies in the city. "In fact, we've received requests to add some artificial noise to the buses so that people can hear them. We're considering it."

The benefits from the switch from diesel buses to electric are not confined to less noise pollution: This fast-growing megacity of 12 million is also expected to achieve an estimated reduction in CO_2 emissions of 48% and cuts in pollutants such as nitrogen oxides, non-methane hydrocarbons and particulate matter.

Shenzhen Bus Group estimates it has been able to conserve 160,000 tonnes of coal per year and reduce annual CO_2 emissions by 440,000 tonnes. Its fuel bill has halved.

"With diesel buses I can remember standing at the bus stop and the heat, noise and emissions they generated made it unbearable in the summer," says Ma. "The electric buses have made a tremendous difference."

China's drive to reduce the choking smog that envelops many of its major cities has propelled a huge investment in electric transport. Although it remains expensive for cities to introduce electric buses— one bus costs around 1.8 million yuan—Shenzhen was able to go all-electric thanks to generous subsidies from both central and local government.

"Typically, more than half of the cost of the bus is subsidized by the government," says Ma. "In terms of operation there is another subsidy: If we run our buses for a distance of more than 60,000 km we receive just under 500,000 yuan from local government."

This subsidy is put towards reducing the cost of the bus fares: "The government looks at the public transport very much as social welfare."

To keep Shenzhen's electric vehicle fleet running, the city has built around 40,000 charging piles. Shenzhen Bus Company has 180 depots with their own charging facilities installed. One of its major depots in Futian can accommodate around 20 buses at the same time.

"Most of the buses we charge overnight for two hours and then they can run their entire service, as the range of the bus is 200 km per charge," says Ma.

There is also geography to consider. Shenzhen is fairly flat, but the hills of nearby Hong Kong have proven too much in trials of electric buses. Other cities in northern China have struggled with battery power in the extreme cold of winter.

Meanwhile, cities such as London and New York are accelerating their drive towards electric buses. London plans to make all single-decker buses emission-free by 2020, and all double-deckers hybrid by 2019. New York plans to make its bus fleet all-electric by 2040.

Viewing

What Happens When You Change Energy Culture?

Video One

Dr. Janet Stephenson (Centre for Sustainability): The average vehicle kilometers traveled by people has actually peaked in the staffing to drop away. So, whereas traditionally agencies who are responsible for investment in transport infrastructure have assumed that vehicle kilometers traveled is going to continue up, therefore we're going to have to keep building more roads and more road infrastructure. Suddenly, all over the world, these curves are starting to flatten off, so there are new changes having to occur within the investment decisions of transport authorities.

Andrew Jackson (Ministry of Transport): Up till around 2004 we saw a rapid increase in the total vehicle kilometers traveled, so the amount of people or driving is going up at about 3% per year. From 2004 to 2012, that was flat. It seems that younger people typically are starting to drive slightly later. There's also a marked decrease in the amount of people who drive when they reach 65 or when they retire. You've also seen economic impacts through the global financial crisis and uncertainty about whether people will have a job.

Dr. Janet Stephenson (Centre for Sustainability): So, energy culture in its most simple way says that what we do doesn't happen in a vacuum. It happens really because we're responding to other things in our lives and importantly what we're responding to is the things that surround us, the physical things that we have in our lives. Also, the norms or expectations that we have about our everyday lives. So, what is it that we think is normal and expected in the way we live our lives? These three things are: our actions or activities, the physical things we have, and how we think where the norms are, which all interact with each other quite strongly to actually build up patterns of behavior that become normalized and habitual in our everyday lives. When we think about mobility cultures, it's the interplay between what we have, what we do, and what our expectations are around our transport.

In New Zealand, we tend to have a mobility culture generally that's really largely fixed around our cars, so you might call the "car culture". But the research that we've done indicates that although people used cars for most of their mobility needs in the everyday lives in New Zealand, there is a bit of a mismatch because a lot of New Zealanders actually have aspirations and they are thinking about that bubble of norms. A lot of people have aspirations to use active transport a whole lot more, so to walk and cycle more than they do or to use public transport more than they do. But somehow, there's this mismatch between what they're actually doing and what their aspirations are. But that mismatch actually creates a bit of opportunity or opening for policy interventions to actually support change in that way so to support people to actually meet their own aspirations.

Andrew Jackson (Ministry of Transport): We shouldn't think of ourselves as a victim to circumstance but rather the captain of our own destiny. We're able to take a decision by the investments that we make, which will affect the transport choices we have. We have a choice ourselves as a nation. A key thing that we want to encourage is a debate on what the nation wants.

Video Two

Andrew Jackson (Ministry of Transport): We've created four different visions of the future of the transport system. We've seen a rapid increase in the numbers of people who are traveling by air. Just like you might use your smart card in Wellington or pop a card in Auckland, you'll have a card like that and just swipe on an airplane. Electric vehicles would become far more common, and we think that the cost of electric vehicles will decrease significantly over the next 10–15 years. Efficiency measures will drive far more efficient combustion engines. It won't be one liter to drive 10 kilometers. It will be 1 liter to drive 100 kilometers. In our vision, we do see that we will see self-driving vehicles; no longer need to regulate whether somebody can drive because people won't drive that will be proven. We've gone from a situation where we use the map to find a way around, to have a sophisticated GPS to being told information of when the bus will arrive. And we think the next evolution will be that we become the managed part of the system and we will be told when there is a slot available in the transport system to get us where we are to where we need to be.

Dr. Janet Stephenson (Centre for Sustainability): Sometimes energy culture has changed when there isn't any obvious single cause. That is really interesting, and we have seen those kinds of changes happening in people's transport lives right now. So, one fascinating example is how quite a lot of young people these days are opting out of owning cars. This change amongst what we call "Generation Y" is really being seen as a change in the culture of young people's mobility, particularly around their daily practices. So, instead of hopping in a car to get somewhere, they tend to use active transport: They walk, they cycle, they use public transport and also they use social media, and of course, they share cars with other people. I guess a whole norm shift here, which is about thinking in different ways about transport. Getting from A to B is not being so much as a problem if it takes time, but (about) actually enjoying the time it takes. So, here's a wonderful example of how mobility culture shifts are actually driving some changes in the general structure around, and making transport authorities and those of others have to think about transport provision in quite new ways.

Andrew Jackson (Ministry of Transport): We valued the research community immensely. We are here to provide advice to ministers on issues. There are often views which were expressed, where people are very passionate about an issue and they will want change, but there are always trade-offs. So, when you're providing advice, one of the key things to have is evidence to allow you to provide advice which will deliver the best outcome for New Zealand.

Dr. Janet Stephenson (Centre for Sustainability): So, the energy culture's framework helps us think about a really complicated situation in a relatively simple way. If people do start changing the energy culture, whether it is because of outside influences or because they are actually changing the norms, wanting to do things quite differently, like "Generation Y" around mobility, then, when people start changing those energy cultures, their "starts" create a collective change. Those collective changes can actually start to influence externally and can start to change the way companies run their businesses; they can start to change the sorts of infrastructure that governments need to provide; they start to change, maybe, some of the policy mechanisms that their governments have. So, that framework, although it's very simple, invites us to think about all of the different parts of this very complex system, and think about both what locks us into stability or habit, and also what it is that drives change.

Unit 6   Energy and Humanity

Listening

Solar Energy Makes the Difference in Africa

Lack of inexpensive, reliable energy delivery is one of the chief obstacles to growth and development in sub-Saharan Africa. Nearly seventy percent of people living in the region lack access to electricity, forcing them to spend significant amounts of their income on costly and unhealthy forms of energy such as diesel fuel to run generators and kerosene for lanterns.

But if there is one thing Africa is not lacking, it is sunshine. And that means a lot; indeed, for some of the world's poorest people, it makes all the difference in the world.

A little more than a year ago, Power Africa—a U.S. Government initiative coordinated by the U.S. Agency for International Development, or USAID, and Power Africa's partners, the United Kingdom Department for International Development, Shell Foundation, and the African Development Bank—launched the Scaling Off-Grid Energy Grand Challenge for Development, with a focus on pay-as-you-go solar home systems. The Challenge's goal is to provide 20 million off-grid households in sub-Saharan Africa with clean, affordable electricity by 2030. So far, the Challenge has made 40-plus investments in early-stage off-grid energy companies, which are expected to result in some 4.8 million new electrical connections.

One reason for the success of the start-up companies that have won grants from the Challenge is the fact that the cost of solar energy is falling fast, so it is within the means of those who live away from traditional power grids. And thanks to the new system of mobile money and pay-as-you-go financing, they can purchase the solar power-generating equipment they need and pay for it in daily installments from their telephones, for as little as 15 cents a day.

Thanks to these technological and financial innovations, many people are getting electricity for the first time, every day. In these newly-electrified communities, businesses can flourish, clinics can safely store vaccines, and students may study long after dark. Indeed, access to clean and reliable electricity can enable entire communities to escape the cycle of extreme poverty.

The Scaling Off-Grid Energy Grand Challenge for Development partnership, solar technology, and companies that bring light and electricity to off-the-grid homes and businesses, are revolutionizing daily life for millions of people in sub-Saharan Africa.

Lack of Access to Energy

For many decades, fossil fuels such as coal, oil or gas have been major sources of electricity production, but burning carbon fuels produces large amounts of greenhouse gases which cause climate change and have harmful impacts on people's well-being and the environment. Moreover, global electricity use is rising rapidly. In a nutshell, without a stable electricity supply, countries will not be able to power their economies.

How Many People Are Living Without Electricity?

Nearly 9 out of 10 people now have access to electricity, but reaching the unserved 789 million around the world—548 million people in sub-Saharan Africa alone—that lack access will require increased efforts.

What Are the Consequences to Lack of Access to Energy?

Without electricity, women and girls have to spend hours

fetching water, clinics cannot store vaccines for children, many schoolchildren cannot do homework at night, and people cannot run competitive businesses. Slow progress towards clean cooking solutions is of grave global concern, affecting both human health and the environment, and if we don't meet our goal by 2030, nearly a third of the world's population—mostly women and children—will continue to be exposed to harmful household air pollution.

Lack of access to energy may hamper efforts to contain COVID-19 across many parts of the world. Energy services are key to preventing disease and fighting pandemics—from powering healthcare facilities and supplying clean water for essential hygiene, to enabling communications and IT services that connect people while maintaining social distancing.

What Can We Do to Fix These Issues?

Countries can accelerate the transition to an affordable, reliable, and sustainable energy system by investing in renewable energy resources, prioritizing energy-efficient practices, and adopting clean energy technologies and infrastructure.

Businesses can maintain and protect ecosystems and commit to sourcing 100% of operational electricity needs from renewable sources.

Employers can reduce the internal demand for transport by prioritizing telecommunications and incentivize less energy intensive modes such as train travel over auto and air travel. Investors can invest more in sustainable energy services, bringing new technologies to the market quickly from a diverse supplier base.

You can save electricity by plugging appliances into a power strip and turning them off completely when not in use, including your computer. You can also bike, walk or take public transport to reduce carbon emissions.

Viewing

Humans and Energy

Video One

So today we're going to talk about *Children of the Sun* by Alfred Crosby who you might remember from our episode on *The Columbian Exchange*. This is Crosby's book about energy and in it, he says: "Modern civilization is the product of an energy binge...but humankind's unappeasable appetite for energy makes the solutions ephemeral and the challenge permanent," which is not that hopeful. But before we start looking at the history of human energy use, let's talk about what we mean by energy.

For our purposes here, energy is the power to do work. For more than 99% of human history, the main source of energy to do work was muscle, either human or animal. And the fuel for that muscle was food, usually plants, and plants ultimately get their energy from the sun. So, almost all the energy that humans use comes from the sun in one way or another. Hence, the book's title is *Children of the Sun*. Humans are a lot of things, but an efficient energy converter isn't one of them. That's why you need a lot of humans to do a lot of work. It took a lot of power to build the pyramids, for example, and it couldn't have happened without some technological advances using energy. The first great energy technology was fire. It enabled us to cook which gave us a greater variety of available food and thus more fuel for our muscles. Fire also led to metalwork and improvements in tools. Another notable advance in energy was the domestication of plants and animals. By domesticating plants, humans redirected the sun's energy to create more nutritious and energy-producing food. The sun also indirectly fueled domesticated animals like horses and oxen, which were harnessed to do even more useful work. After the invention of agriculture, developments in human energy kind of plateaued for a while. The only energy that we had that didn't derive from the sun was water power. Since wind technically comes from the sun's heating the air, sailing ships and windmills are kind of solar power. There were some minor advances, like concentrating the energy density of wood by converting it to charcoal and adding oxygen to fires using bellows, but for the most part, power was still generated by muscle.

The next big change in energy use came with industrialization. Let's go to The Thought Bubble. So, industrialization utilized new forms of fuel in coal, and later oil and natural gas. These fuels are just really, really old forms of fossilized plant and animal matter. So again, they're originally from the sun, but we don't think of it this way. Nobody calls coal "solar power". While the Chinese were using coal during the Song Dynasty to work iron, for example, England was where coal use really took off, thanks to the steam engine. Newcomen's steam engine was, according to Crosby, "the first machine to provide significantly large amounts of power not derived from muscle, water, or wind". Coal-powered industrialization was a pretty big deal. It allowed Britain to dominate the textile industry, industrially produce weapons and steam-powered ships and enable Europeans to penetrate and dominate Africa and parts of Asia. According to Crosby, fossil fuels "created the political and economic landscape we recognize today".

Video Two

After steam-powered manufacturing, it was a short chronological leap to electricity, which was used to power machines, and for illumination. Electric light was a really big deal because it provided a clean and efficient way to allow people to work after dark. Thanks, Thought Bubble! Although we might think of coal as the fuel of the 19th century, we still use a lot of it today, especially for generating electricity, but coal is much less efficient than oil. Oil was revolutionary because it could power not only electricity plants, ships, and trains, but also the internal combustion engine, which makes cars and trucks possible. Crosby maintains that "the internal combustion engine powering the automobile, truck, and tractor has for a century been the most influential contrivance on the planet". By the end of the 20th century, there were half a billion cars in the world, and humans were using 70 million barrels of oil each day. Manufacturing and driving all those cars has had a huge impact on the environment. From an energy use perspective, the world since 1900 is a totally new era in human history. We use electricity for everything in the West: It powers our gadgets; it lights our homes; it gets us around on trains, or buses, or cars. Crosby puts it like this: "Humanity's primary energy use has increased twenty times over since 1850 and nearly five times over since 1950. In the U.S., each individual consumed 2,000 kilowatt-hours of electricity in 1950, and 32,700 in 2000. Oil and natural gas are the most important fuels for this electricity boom, although as of 2006, 40%–50% of humans, most of them living in the tropics, still rely on wood for fuel."

So, "When the world will run out of oil?" is a topic of heated debate. But scientists have been looking for other forms of fuel for decades. One alternative is nuclear power, which has not been a total success. The first nuclear plant providing power for homes opened in the Soviet Union in 1954, and some countries, notably France, still rely heavily on nuclear energy. Despite initial enthusiasm from scientists and science fiction writers, nuclear power never caught on in the U.S. partly because it's really expensive. Another problem is that no one can figure out what to do with the radioactive waste that nuclear energy produces. But the biggest reason nuclear power fails to catch on is that people think nuclear power is dangerous, believing that nuclear plants can easily turn into huge bombs. Nuclear accidents have happened, though, notably Windscale in England in 1957, and San Loren in France in 1969, but neither of these was catastrophic. The U.S. had a nuclear scare in 1979 with an accident at Three Mile Island in Pennsylvania. Although there were no immediate casualties, thousands of people in the vicinity were forced to evacuate, and the cleanup took years, and cost millions. The disaster at the Soviet nuclear plant at Chernobyl in 1986 was much worse, with a release of radiation that was hundreds of times greater than that given off at Hiroshima and Nagasaki, and the fallout that will be lethal for 24,000 years. A few countries still use nuclear power, but it never really caught on.

Overall, nuclear power has never accounted for more than 5% of the world's energy supply. In recent years, rising concerns over climate change have led to increased calls for humanity to find cleaner, more renewable forms of energy. The alternative—that we significantly reduce our energy consumption—seems unlikely. Especially since it would seem like a historical step backward. History is often presented as a story of progress and growth and increasing complexity and a future in which we use less energy is kind of hard to imagine. But historically speaking, the world we live in is new; it's unsustainable, and it's not normal. Crosby offers us this reminder: "Most of us in the richer societies can only recall times of immediate access to abundant energy. That abundance tempts us, successfully, to believe that having energy flow down lines from far away and illuminate our rooms when we flip the switch is normal rather than miraculous." In the end, how we reconcile our desire to continue our history of growth and rising complexity with the fact that such growth is unsustainable with current technology is one of the biggest challenges facing humanity today. Whether and how we rise to that challenge will determine what kind of world we live in tomorrow.

Unit 7   Energy and Environment

Listening

Tons of Methane Gas Could Be Trapped Under Antarctica

An international team of scientists has found that up to four billion tons of methane gas could be trapped under ice-covered areas of Antarctica. The scientists say extremely small organisms may have changed ancient organic matter into methane. And they say if enough ice melts, it is possible that enough of the gas could escape and add to the warming of Earth's atmosphere.

The continent of Antarctica is mainly over the South Pole, almost totally south of the Antarctic Circle. The high pressure and cold temperatures under the ice make good conditions for forming methane hydrate. The scientists say the organic material came from a period thirty five million years ago. At that time, Antarctica was much warmer than today and contained life forms. A researcher from the University of California at Santa Barbara described how methane formed. She said some of the organic material became trapped in material that had fallen to the bottom. These sediments, in her words, were then "cut off from the rest of the world when the ice sheet grew".

The scientists say fifty percent of the West Antarctic ice sheet and twenty-five percent of the East Antarctic sheet are on sedimentary basins. They said these areas hold about twenty one billion tons of carbon.

The Antarctic ice sheet covers land, but not the surrounding sea. Methane hydrates are also found at the bottom of oceans.

Earlier this year, news about methane gas came from the opposite end of the world. Researchers reported finding thousands of places in the Arctic where the gas was rising to the surface.

Studies on the ground and from the air found one hundred fifty thousand seeps—places in Alaska and Greenland where methane had escaped. The seeps were found in lakes along the edges of ice cover.

The study appeared in the journal *Nature Geoscience*. It showed that some of the seeps are freeing ancient methane. The source may be natural gas or coal deposits beneath the lakes.

Other seeps release more recently formed gas. The newer methane may have formed when plant material in the lakes broke down into basic elements.

Another scientist involved in methane research described the Arctic area as the fastest-warming area on Earth. Euan Nisbet from the University of London said the area has many methane sources that will increase as the temperature rises. He says there is a serious concern that the warming could cause more warming.

Some scientists believe the effects of the freeing of methane will not be seen for many years. Others note that the possibility of a fast methane release could speed up rising temperatures on Earth.

Solar Plant Raises Environment Concerns

The world's largest solar thermal plant is set to begin producing power in the United States by the end of the year. Wind and energy from the sun are generally considered clean, unlike energy from coal-burning power stations. However, environmentalists now worry that too much solar power development could harm the local environment.

A California company—BrightSource Energy is building a huge solar power plant in the Mojave desert, about 60 kilometers southwest of Las Vegas, Nevada. The plant is known as the Ivanpah Solar Electric Generating System. Joe Desmond works for the company.

"This is actually one of the highest concentrations of sunlight in the world, out here in Ivanpah," explained Desmond.

BrightSource Energy will deploy 170,000 specially designed mirrors to direct solar energy towards boilers on top of three power towers. The steam produced in the boilers will drive turbines to make electricity. Joe Desmond says the steam can reach temperatures of more than 260 degrees Celsius.

"We can store the sun's thermal energy in the form of molten salt, so we can produce electricity even when the sun goes down. There is a lot of interest in concentrating solar power around the globe in environments where you have lots of sun, such as China, South Africa, the Middle East, North Africa," explained Desmond.

Environmentalists generally support the idea of solar power, however, many are concerned about the effect of power plants on sensitive environment.

Lisa Belenky is a lawyer with the Center for Biological Diversity, a private group. She says environmentalists are specifically worried about the effect of the Ivanpah Solar Project on the sensitive plant and animal life in that part of Mojave desert.

"Even though the desert seems big, when you start cutting it up, it can really affect how the species and the animals and the plants are able to survive in the long run," said Lisa Belenky.

BrightSource Energy has already spent more than $50 million to move endangered desert tortoises away from the power plant, but Lisa Belenky says this is not the answer.

"We should be reusing areas that have already been disturbed, like old mining sites, for example...either on homes, on businesses, parking lots," said Belenky.

There have also been reports of birds dying at the Ivanpah Plant and others like it. Some birds die after colliding with solar equipment which the animals mistake for water. Other birds were killed or suffered burns after flying through the intense heat at the solar thermal plant. As solar projects increase, environmentalists and developers are considering what to do to reduce bird death.

Viewing

Energy and Environment

Video One

In this video, we explore energy and the environment. You'll see why fossil fuels our history, and how we'll provide clean energy and water for the entire planet, and how we'll clean up the toxic waste we've accumulated too. So, let's get started.

Where are we going to get the energy to do all the stuff? Well, let's talk about energy and environmental remediation. What we're recognizing is the toxic sludge is made from atoms. So, let's rearrange the atoms and turn toxic sludge into a harmless inert chemical. In Japan's landfills, they have a 20-year supply of gold and other materials, enough to meet the needs of the entire world. So, let's transform waste into value. Take a look at the smog-free project going underway. Launched in 2015 in Rotterdam, it's the largest air purifier in the world. It's a tower twenty-three feet tall. It collects the pollution from the air and it not only removes it. It compresses it into tiny cubes that they use to make jewelry out of. How cool is that! There are more towers planned in cities around the world.

How about energy creation? Well, forget about fossil fuels. That's so 20th century. It's the sun that's going to be the key provider of energy for our planet. The sun produces 10,000 times of our needs on a global basis. So in other words, it produces way more energy than we will ever be able to use. In the deserts of North Africa, solar power can supply 40 times the world's demand—a square meter of Sun in the desert is the same as one and a half million barrels of oil and 300 tons of coal. The cost of solar is dropping. This is why you didn't see solar, and still don't today see solar as a big deal because it's so expensive. In 1984, it was $30 a kilowatt-hour; by 2014, was down to 16 cents, a 200x improvement in 20 years. And it's continuing to get cheaper thanks to Moore's Law. Look what's going on at a major airport in India—one and a half million square feet of terminals, 2.3 million passengers a year, 45 acres of solar panels. Look at the terminal buildings. They're there. What you're looking at in the foreground are the solar panels. The terminal buildings in the runway are smaller, much smaller in comparison. This is going to cut 300,000 tons of carbon emissions over 25 years; the cost of solar is dropping 50% every 24 months. By 2040, we will be producing 100% of the world's needs from solar; by 2042, it'll be 200% because the cost is dropping 50% every 24 months; by 2044, it'll be 400%; by 2046, 800%. This again is the power of compounding.

It's not enough to produce the energy. You then have to store it for future use. Well, that means battery technology is vital. Batteries of the future will be using sodium and water instead of lithium, releasing energy evenly, safe enough to eat, cheap, quiet, no maintenance, lasting decades. The price of photovoltaic cells is dropping 60% since 2001. Soon it'll be cheaper than oil even if oil is ten bucks a barrel. Developers at MIT and Samsung have built an almost perfect battery: rechargeable, never wears out, can't overheat, 30% smaller than current batteries. Pretty exciting stuff isn't it? Not if you're in the power business. Edison Electric Institute says utilities face the risk of declining retail sales, loss of customers, potential obsolescence.

Video Two

Take a look at that building in New York. A single 30-story building could feed 50,000 people a year. This is already in place, there's an old steel factory in Newark, New Jersey. It's now the world's largest vertical farm. It grows over 200,000 pounds of produce a year. That's 75 times more productive than a traditional farm of similar size: it recycles materials; it uses a lighting system; it uses less energy; it requires no soil—needs only 5% as much water as traditional farms; and it's pesticide-free. The mist is packed with nutrients and oxygen; the full crop cycle is just 16 days.

And here is one of my favorites—the toilets. Toilets use 31% of all the water in the U.S. A 1.25-gallon leak occurs from every toilet every year in the United States. It's the nation's biggest water waster. The Bill and Melinda Gates Foundation is working on a solution. Their goal is to improve the age-old toilet. It's the 21st century toilet and here's a prototype. It's amazing how it works. Let me explain to you why this is such a big deal. You know, you say it yourself: Why do we need a new and improved toilet? Okay, we got water waste and so on. But it's not, it's a much bigger deal in Africa, and here's why. You've got villages in Africa where there is no electricity, there's no running water, no paved roads. These folks are living in huts, and they have no heat and they have no electricity. How do they cook? How do they heat their huts? Well, the children go off into the fields looking for animal dung. This takes hours. It prevents the children from going to school: They collect the dung; they bring it back to the hut. Mom uses it to cook, sets the dung on fire, and uses it to cook and to heat the hut. There's one big problem though. Burning animal dung is extraordinarily toxic, and more people die in Africa of respiratory disease than any other single cause. What choice do they have? Well, here's a choice. We bring in the new 21st century toilet. It is completely self-contained—it has no pipes, and no septic system, no sewers of any kind. The system burns the feces and flash evaporates the urine. Everything is sterilized. The result is that you now have fertilizer for growing crops, table salt, freshwater, and enough electricity to power the home for about a nickel a day. It reduces disease and now creates economic freedom: The children don't have to spend hours looking for fuel, they can go to school; the parents can go to work. Imagine the impact on the rising billions— the billions of our planet who are at the lowest level of the economic spectrum, who live on a dollar a day. Imagine the impact on them, as a result, of a 21st century toilet.

So what are the personal finance implications of all these changes in energy and the environment? Well, you can expect energy costs to continue dropping, but you're going to want to refurbish your home, to take advantage of these solar panels and the likes, and that will be expensive. Number two, I don't see any real reason to be buying specifically utility stocks these days, because utilities which are using fossil fuels, nuclear energy and other sources, are fast becoming obsolete. And obsolete is what we want to help you avoid as we explore the truth about your future.

Unit 8 Energy and Future

Listening

Islay's Green Energy Projects for the Future

A Scottish island could become one of the first places to become self-sufficient in renewable energy. The centerpiece of Islay's plan is a community-owned Wind Farm, which will provide around half of the island's energy needs. But the island is also storming ahead with a variety of other green energy projects.

Harnessing the power of the wind, the sea, the sun and the earth—that's the dream of one Scottish island. Under a scheme sponsored by the European Commission, Islay's 3,000 inhabitants intend to become self-sufficient in renewable energy, and to become a model to the world. Islay already has the world's only commercially operational wave power station. Its designer, Tom Heath, says the potential energy in the sea is almost limitless. Islay's wave power station is one of the reasons the European Commission chose the island for its RESPIRE Project. It aims to demonstrate the feasibility of island communities going totally green. The next stage is to get planning permission to build three wind turbines—the centerpiece of the RESPIRE Project, supplying half of Islay's energy needs. Central to the whole project is the idea that, once the Wind Farm is operational, it'll be run by and for the people of Islay, through a community-owned energy company.

Islay is forging ahead with other green energy projects. One of its world-famous whiskey distilleries uses waste hot water to heat the local swimming pool. Islay has a wave-powered bus—its batteries are charged with electricity produced by the wave station. New houses are being built with geothermal heating, tapping into warmth fifty meters below the Earth's surface. And, even in a climate where there's more rain than sun, this Gaelic Language Center is powered by solar panels.

Achieving a hundred percent self-sufficiency in renewable energy may be an unrealistic target. But even if it only gets close, Islay could be a model for other islands around Europe, looking for ways to help both the environment and their own communities.

Scottish Whisky Tested as Alternative to Fossil Fuels

Waste products from a popular alcoholic drink could be used in the future to make biofuel. Researchers say the new fuel, based on whisky, could reduce demand for oil. They say using less oil could cut pollution that studies have linked to climate change.

Scotland is the largest producer of whisky in the world. And a Scottish professor has found how to take the byproducts from distilling whisky and turn them into a form of alcohol called biobutanol. Biobutanol can be used as fuel.

Whisky comes from grain, such as corn, rye or wheat. First, the liquid is purified. It is heated until it becomes a gas. The gas is then cooled, and the resulting liquid collected.

A research center says less than 10 percent of what comes out can be considered whisky. The rest is mainly one of two unwanted byproducts: pot ale and barley. The two byproducts can be produced to create a new material: biobutanol. The whisky-based biofuel provides more power than bioethanol, a fuel made from plants like corn or sugar cane. It has almost the same amount of energy as the gasoline used to power automobiles. The research center also says biobutanol is not expected to replace gasoline, but the two fuels can be mixed together. It is possible that the fuel may also be used in airplanes, ships and in heaters. Drivers would also be helping the environment by "reducing the oil that we consume by putting this into their cars".

A company called Celtic Renewables has received $17 million from the British government to build a center producing biobutanol in Scotland. The center is expected to be operational within three years.

Viewing

Hydrogen—the Fuel of the Future

Video One

I think we can all agree the sooner we decrease our reliance on fossil fuels and develop new energy sources, the better, whether you believe in climate change or not, the benefits extend beyond just the reduction in greenhouse gas emissions and the supply of oil and gas will inevitably dry. Tesla pioneered our greatest hope in this space to date with the development and popularization of battery technology. But as we've seen, they are struggling to meet the enormous half a million pre-orders for the Model 3. Elon self-proclaimed production hell has resulted in delay after delay. Bloomberg estimates that Tesla have produced around 12,000 Tesla Model 3 to date, with the current production rate of 1,000 per week which will gradually grow to a target output of 5,000 per week. This is just the tip of the iceberg.

Last year, 72 million passengers carriers were built, that's nearly 1.4 million vehicles a week. No matter how successful the Internet wants Tesla to become, they will never solve this issue alone. And the industry as a whole, likely won't be able to solve it with a battery-only approach. The demand for lithium-ion battery technology is simply growing faster than the supply of lithium can satisfy. So, it seems clear. We need a multi-faceted approach to solve this problem. Another solution which was the industry favored to take over from fossil fuels, not so long ago, is hydrogen fuel technology, and companies like Toyota and Shell are working to develop this industry. It won't be an easy race, but hydrogen may well prove to be the tortoise that beats the hare.

Hydrogen has three primary obstacles that needs to overcome to become a viable energy source for any industry: safety, infrastructure and cost. Let's get the big elephant in the room out of the way first, I know it's on your mind. If hydrogen fuel cells are ever going to make it to public roads at scale, the hydrogen needs not only to be safe, but to be perceived as safe. And, yes, filling a gigantic incredibly flammable balloon with hydrogen is a pretty bad idea. Hydrogen has a relatively low ignition temperature and a very wide ignition range for air to fuel mixture percentages. The fact that it's pressurized makes explosions a worry, but it has one massive advantage over oil derived fuels: It's lighter than air. It can be purged using emergency valves in the event of a fire and if it does ignite it won't pull around the vehicle, engulfing it and its passengers in flames. Toyota even tested their carbon fiber tank by shooting it with a 50 caliber round, the tank didn't explode, it simply let the lighter-than-air gas to escape and vent to atmosphere. Hydrogen is arguably safer than gasoline, so safety isn't a huge concern for hydrogen, but the lack of infrastructure is. Battery-operated vehicles have had a huge head start in this space. The electric grid is a prebuilt transportation and generation network for the fuel the battery-operated vehicles require, and installing a charger in your driveway or garage isn't a huge challenge. Hydrogen doesn't have such luxuries to kick-start the hydrogen economy.

There are a few large scale production facilities in the world, with the largest being Shell's Rhineland oil refinement facility. It uses its own hydrogen production in the oil refinement process, but the lessons learned from these efforts have allowed Shell and its partner ITM to make hydrogen a viable option for uses in energy storage. Last month, I was invited to London to witness the opening of the UK's first ever hydrogen fuel pump, to be included under a fuel station canopy—a pivotal step in making the public see hydrogen as an integral part of the transport ecosystem. What fascinated me about this site was how the hydrogen got there. Transporting hydrogen in pressurized trucks would be too expensive as there are no large-scale production facilities nearby. Although hydrogen can be transported within the already established natural gas pipelines around the world, for use in vehicles we need pure hydrogen. So, Shell and ITM took the next logical step to keep cost down. They built a hydrogen production and storage facility on site. The production facility is placed just behind the main station and is capable of producing 80 kilograms of hydrogen a day. The Toyota Mirai is on site, it had a range of 480 kilometers with a full 5 kilogram tank of hydrogen, vastly more than a full charge for a Tesla, but you must consider the huge upfront cost of batteries which do not last forever in this equation for cost.

Video Two

I'll explore this battery versus hydrogen dilemma more in a future video but for now let's see how hydrogen actually works.

The production process of hydrogen is pretty simple. It uses a process called electrolysis to separate water into hydrogen and oxygen. The electrolyser consists of two metal coated electrodes and a DC power source which provides a negative and positive charge. Hydrogen will appear at the cathode, the negative electrode, where electrons react with the water to form hydrogen and hydroxide ions. These negative ions now present in the water are attracted to the anode or positive electrode where they are oxidized to form oxygen and water. The rate of production of oxygen and hydrogen depend on the electric current, but pure water is not very conductive. To achieve adequate hydrogen production, we would need to increase the voltage or increase the conductivity. It's much more efficient to increase conductivity, so an electrolyte in the form of salt is often included as a charge carrier. This is the oldest and most well-established production method for hydrogen. For reasons, I won't go into but will include reading materials in the description; this method isn't suitable for quick response times with slow starter procedures and safety concerns, making it completely unsuitable for variable renewable energy sources which have historically made hydrogen prohibitively expensive.

If hydrogen has any hope of becoming a popular fuel source, we first need to get its price down to be competitive with batteries and fossil fuels. This has been a major point of research for the past 50 years, and PEM or proton exchange membranes as the primary solution now come into market and are facilitating a realistic hydrogen economy. PEM replaces the electrolyte-rich water for a solid polymer electrolyte membrane, sandwiched between the anode and cathode, with channels to allow water and gas and solution to flow through. As its name suggests, the PEM only allows protons to pass through, so hydrogen ions, otherwise known as protons, now become the charge carriers rather than a hydroxide ions. But the overall chemical reaction is exactly the same while requiring less voltage to operate efficiently and more importantly has a rapid response time, making it ideal for integration to the grid as an energy storage method. This is where it truly dries on costs. The hydrogen fuel cells and carriers use this exact process in reverse to power their electric motors. The cost of hydrogen production by electrolysis is completely dependent on electricity prices. If an electrolyser cannot take advantage of cheaper intermittent surge electricity or use cheaper off-peak electricity, then it's losing out on real cost savings and can't provide the valuable service of energy storage for the grid.

But this hydrogen facility at the Shell station can form an important part of the renewable grid infrastructure going forward. Hydrogen's greatest chance at success is by fueling a new economy of hydrogen, where natural gas pipelines are supplemented with hydrogen produced with cheap renewable energy, allowing hydrogen to gradually grow to be the Earth's primary energy storage method and facilitating renewable energy to become a larger part of our energy grid without the worry of weather impacting energy supply, allowing nations to stop depending on the importation of fossil fuels, and instead grow their own fuel economy.

One tiny group of isolated islands in the Bay of my home county of Galway is attempting to do just this. The Aran Islands are rural Irish-speaking islands—popular resorts for their unique landscape—that would have historically depended completely on the mainland for fuel. There are no trees here, no coal, no turf, no oil, but what they do have in plentiful supply is wave and wind energy. They are the perfect candidates to develop a mini hydrogen economy—an economy where they generate their own renewable energy and create their own field to heat their homes and power their vehicles. Who knows, these tiny obscure Irish islands could be the birthplace of the world's first self-sustained, renewable, zero-carbon hydrogen economy.

教师服务

感谢您选用清华大学出版社的教材！为了更好地服务教学，我们为授课教师提供本学科重点教材信息及样书，请您扫码获取。

》最新书目

扫码获取 2024 **外语类**重点教材信息

》样书赠送

教师扫码即可获取样书